CAMBRIDGE MONOGRAPHS ON PHYSICS

NUCLEAR STABILITY RULES

NUCLEAR STABILITY RULES

BY

N. FEATHER, F.R.S.

Professor of Natural Philosophy in the University of Edinburgh

CAMBRIDGE
AT THE UNIVERSITY PRESS
1952

CAMBRIDGE
UNIVERSITY PRESS

University Printing House, Cambridge CB2 8BS, United Kingdom

Cambridge University Press is part of the University of Cambridge.

It furthers the University's mission by disseminating knowledge in the pursuit of education, learning and research at the highest international levels of excellence.

www.cambridge.org
Information on this title: www.cambridge.org/9781316601846

First published 1952
First paperback edition 2015

A catalogue record for this publication is available from the British Library

ISBN 978-1-316-60184-6 Paperback

GENERAL PREFACE

The Cambridge Physical Tracts, out of which this series of Monographs has developed, were planned and originally published in a period when book production was a fairly rapid process. Unfortunately, that is no longer so, and to meet the new situation a change of title and a slight change of emphasis have been decided on. The major aim of the series will still be the presentation of the results of recent research, but individual volumes will be somewhat more substantial, and more comprehensive in scope, than were the volumes of the older series. This will be true, in many cases, of new editions of the Tracts, as these are re-published in the expanded series, and it will be true in most cases of the Monographs which have been written since the War or are still to be written.

The aim will be that the series as a whole shall remain representative of the entire field of pure physics, but it will occasion no surprise if, during the next few years, the subject of nuclear physics claims a large share of attention. Only in this way can justice be done to the enormous advances in this field of research over the War years.

N. F.
D. S.

CONTENTS

CONTENTS

AUTHOR'S PREFACE

This short monograph has been more than five years in preparation, and much of it has been re-written several times during that period. Its origins date back even further: the germ of Chapter III, § 3, is to be found in a British Atomic Energy Report written in April 1943, and the hard core of Chapter II in another report, of the same series, dated August 1945. Chapter I was completed in first draft in 1946. Only Chapters III and IV, in their present form, are the result of fairly continuous writing. In such circumstances it cannot be expected that the attempt to bring the whole work up to date, as at the time of delivery to the printer, could be uniformly successful. Nevertheless the attempt has been made, the relevant date being mid-December 1951.

More difficult than the continual revision of experimental results quoted in the text (or incorporated in the diagrams) has been the maintenance of a steady aim in presentation, during a time in which the subject has developed with such amazing rapidity. The more reason, then, that that aim should be plainly confessed, now that the writing is done. It has been, throughout, to survey the results of experiment, interpreted as providing information concerning the stability properties of the normal (ground) states of nuclei, with a view to eliciting the significant regularities. Pursued literally and exclusively, this aim would issue in an impossibly dreary performance; obviously, the exercise in pure empiricism must be relieved by an assessment of significance in terms of current 'theoretical' ideas—if it is to be rewarding, even if it is to be comprehensible. But it cannot too often be emphasized that, in so far as they are free from unsuspected error, the experimental results are unalterable; the theoretical ideas are not. Stated crudely, the aim of this monograph has been so to marshal the experimental facts that the theorist is most likely to be inspired by valid—even 'correct'—ideas on contemplating them. If there is any who will say that the theorist has no need of the facts in order to be inspired by a valid theory, this monograph is not addressed to him.

NORMAN FEATHER

EDINBURGH
3 *January* 1952

CHAPTER I

THE SYSTEMATICS OF STABLE NUCLEI

1.1 Empirical definition of stability

The concept of nuclear stability requires careful definition. Formally the matter is simple enough: a nucleus is said to be stable when it is not subject to spontaneous disintegration; in practice, however, it is not always easy to decide the question one way or the other by direct experiment.

For nearly forty years α- and β-disintegration were the only types of spontaneous disintegration known to occur, but the study of 'artificially' produced species during less than half that time has added positron emission and orbital electron capture to this list—and the more recent discovery of neutron-induced fission has been followed by the recognition that fission also occurs as a spontaneous process with the heaviest nuclei (Petrzhak and Flerov, 1940). Here we may note a problem in conventional nomenclature. We have to envisage the possibility of types of spontaneous transformation in which nuclei divide into two fragments between which the mass is shared in any proportion: obviously the term fission might serve to describe all such processes (less well, admittedly, for electron and positron emission) but it has acquired a particular meaning. In actual fact the acute problem has not yet arisen; we speak of α-disintegration when the lighter fragment is the helium nucleus, we should certainly speak of spontaneous proton emission if a hydrogen nucleus were involved, but we have also to be prepared in theory to consider other light nuclei as possible products of transformation. It may conceivably be held that the characteristic of 'fission' is that any spontaneously fissionable nucleus may divide in a number of different ways, but in the limiting case clearly this number will not be greater than one, and the distinction will disappear. Evidently there is a problem in nomenclature involved, though it is as yet an academic one.

It has been implied that it is not always easy to decide by direct experiment whether a given element is stable or unstable, that is whether or not any isotopic constituent of the element is unstable as it occurs under terrestrial conditions. Since the present chapter deals with the 'stable' nuclei, defined for our immediate purpose as those nuclei not known to be unstable, it will be well to examine this statement further so that the degree of uncertainty in any case can be appreciated. We shall see that what is important, for all possible types of instability, is the dependence of disintegration probability upon available energy—and also, in certain circumstances, the rate of loss of energy with distance, for the disintegration particles, in the medium in which they are to be detected.

In respect of α-disintegration we may note that no case is known, with the 'classical' radioelements, in which less than 4 MeV. of energy is liberated. This result may well reflect a definite discontinuity in nuclear properties, considered as a function of nuclear charge, for a charge number in the neighbourhood of 82 (see p. 46), but it certainly bears witness to a practical limitation which is undoubtedly effective. The half-value period of thorium (α-disintegration energy 4·05 MeV.) is $1{\cdot}40 \times 10^{10}$ years; if the α-disintegration energy had been 3·3 MeV. or less the half-value period would have been at least 10^{15} years. In that case the natural radioactivity of thorium would probably have escaped recognition hitherto. Similarly, it can be stated that it is impracticable at present to decide, at least by direct experiment, whether or not naturally occurring lead or bismuth are α-active, except in respect of disintegration energies greater than about 3 MeV.† For disintegration energies less than this amount the rate of disintegration of these elements would be too slow for certain detection. For elements of smaller nuclear charge, however, less uncertainty in respect of energy remains; disintegration probabilities increase, for a given energy, as the nuclear charge decreases (p. 74) (the active isotope of samarium, $^{147}_{62}Sm$, has a half-value period of $1{\cdot}1 \times 10^{11}$ years (Cuer and Lattes, 1946; Picciotto, 1949) and a disintegration energy of 2·24 MeV. (Jesse and Sadauskis, 1949)), on the other hand the difficulty of detecting α-particles of small energy emitted by feebly active preparations eventually asserts itself, so that the situation in respect of minimum detectable instability does not

† Faraggi and Berthelot (1951) have recently reported the observation of α-particles of energy 3·15 MeV. which they attribute to the disintegration of $^{209}_{83}Bi$ with half-value period $2{\cdot}7 \times 10^{17}$ years.

improve indefinitely as the atomic number of the element concerned is reduced to the limit. Direct experiment has been unable to establish the stability of the common isotope of beryllium ($^9_4$Be) beyond a possible half-value period of 10^{14} years, assuming that the energy of α-emission might be as small as 0·1 MeV.† Probably the most sensitive method of investigation of general applicability is the method which employs the newly developed fine-grain emulsions for the detection of the emitted particles, but even with these emulsions it is difficult to recognize with certainty the tracks of α-particles of less than about 0·6 MeV. energy, that is, of less than 4 mm. range in standard air. As recently employed, this method has yielded minimum lifetimes of the order of 10^{16} years for certain medium-heavy elements.‡

The problem of feeble β-activities is not complicated by any important variation of lifetime with nuclear charge, for a given energy of disintegration. In another respect, however, the position is more complicated than with α-disintegration. In the case of β-disintegration, disintegration probabilities are not represented by a single-valued function of the energy, as, to a first approximation, α-disintegration probabilities are, for a specified nuclear charge. Depending upon the details of the transformation (cf. p. 93), β-disintegrations are classified as 'allowed', 'once-forbidden', 'twice-forbidden', etc., and disintegration probabilities may vary by a factor of 10^2 or 10^3, for a given energy, for a change in type of disintegration represented by the difference between one class and the next. If it were not for this effect there would, in fact, be no outstanding problem concerning possible β-activities of long life-time; if all disintegrations were 'allowed' the half-value period for any activity having a characteristic energy of greater than, say, 0·01 MeV. would be measured in thousands of years, at the most. Clearly, no species of such short lifetime could exist in appreciable concentration on the earth to-day, unless it were continuously being produced from a much longer-lived parent which then of necessity would be heavy-particle active. However, all β-disintegrations are not of the 'allowed' type, and within recent years β-activity

† Rayleigh, 1933; Evans and Henderson, 1933; Gans, Harkins and Newson, 1933; Libby, 1934.

‡ Broda, 1947; Jenkner and Broda, 1949; Bestenreiner and Broda, 1949*a*, 1949*b*.

of long lifetime has been established for $^{115}_{49}$In (Martell and Libby, 1950), $^{176}_{71}$Lu (Heyden and Wefelmeier, 1938; Libby, 1939) and $^{187}_{75}$Re (Naldrett and Libby, 1948*a*, 1948*b*; Sugarman and Richter, 1948), and the previously recognized β-activities of potassium and rubidium have been assigned to the common species $^{40}_{19}$K (v. Hevesy, 1935; Smythe and Hemmendinger, 1937) and $^{87}_{37}$Rb (Mattauch, 1937; Hahn, Strassmann and Walling, 1937). Moreover, there have been reports that $^{145}_{60}$Nd† is also β-active (see, however, p. 17), and the recently discovered $^{138}_{57}$La (Inghram, Hayden and Hess, 1947*b*, 1947*c*) and $^{50}_{23}$V (Hess and Inghram, 1949; Leland, 1949*b*) can hardly be β-stable (Pringle, Standil and Roulston, 1950; Pringle, Standil, Taylor and Fryer, 1951). As an indication of the degree of uncertainty in respect of energies and lifetimes which may still persist in respect of a small number of species, possibly subject to highly forbidden β-disintegrations such as these, it may be stated that the maximum energy of the β-particles emitted by $^{187}_{75}$Re is 0·043 MeV. and that the half-value period of this species (of natural abundance 62·9% in ordinary rhenium) is roughly 4×10^{12} years. It is not impossible, therefore, that one or two comparatively rare species, at present classified as stable, may be undetected β-emitters of small disintegration energy—say 0·01 MeV. or less—and of lifetime no greater than 10^9 years.

What has been said concerning β-disintegration applies equally to positron emission, though, up to date, no naturally occurring positron emitter has been discovered. The case of orbital electron capture, however, is less hypothetical, and it raises new problems. It has been established that $^{176}_{71}$Lu undergoes transformation of this type as an alternative to β-disintegration (Flammersfeld, 1947) and the indications are that $^{40}_{19}$K exhibits similar branching,‡ and that $^{138}_{57}$La also is capture-active (vide supra). From the point of view of experiment the main problem posed by the occurrence of orbital electron capture is that the primary radiation is not a particle radiation. Methods of detection, therefore, are generally insensitive. Moreover, with the lighter elements the primary radiation is

† Libby, 1934; Feather, 1945; Ballou, 1948; Jha, 1950; Wapstra, 1952.

‡ v. Weizsäcker, 1937; Thompson and Rowlands, 1943; Bleuler and Gabriel, 1947; Stout, 1949; Sawyer and Wiedenbeck, 1949, 1950; Bell and Cassidy, 1950; Spiers, 1950; Ceccarelli, Quareni and Rostagni, 1950; Inghram, Brown, Patterson and Hess, 1950; Good, 1951; Colgate, 1951.

relatively easily absorbed in the source material—or in the air: the K X-radiation from potassium, for example, is almost completely absorbed in a few centimetres of standard air. It is true that a nuclear γ-radiation may follow the emission of the 'primary' fluorescent X-radiation in an appreciable fraction of the transformations with some capture-active species, making possible the recognition of a number of activities which would otherwise go undetected,† but it is not possible to rely on such a favourable circumstance in all cases. When this feature is absent it can almost be said that any of the lighter capture-active species which is sufficiently long-lived to exist in detectable amount upon the earth is also so feebly active as to appear stable under direct experimental examination. Even with the heaviest species, for capture activity to be recognized with certainty, the lifetime should not be greater than 10^{12} years. Further it must be noted that capture processes may remain undetected even through the energy of instability is not very small: if there is no available state of excitation of the product nucleus (which might be exhibited in γ-ray emission) the whole of the excess energy is carried by the unobservable neutrino.

One final remark concerning limits of detection is called for here. The figures which have so far been given refer to the recognition of instability by the detection of the radiation emitted from disintegrating nuclei. In certain cases geochemical methods provide a more sensitive test. Such methods depend upon the accumulation of a stable end-product of disintegration during geological time, and they are particularly suitable when this end-product is an inert gas. An extreme case is represented by the mass analysis of the xenon occluded in an old tellurium mineral which has established the stability of ${}^{130}_{52}\mathrm{Te}$ against the simultaneous emission of two β-particles as far as a lower limiting lifetime of 8×10^{19} years (Inghram and Reynolds, 1949). Obviously this conclusion depends upon a knowledge of the age of the mineral concerned, and upon the assumption of complete retention of occluded gas during the time for which the mineral has existed in the solid phase. A less extreme case, but one less open to doubt concerning possible disappearance of the end-product of disintegration, is provided by the chemical

† The K-capture transformation of the artificially produced ${}^{7}_{4}\mathrm{Be}$ is an extreme example of this.

estimation of indium in an old cassiterite, which has led to the conclusion that $^{115}_{50}Sn$ is stable against capture transformation unless the lifetime for this process is greater than 5×10^{12} years (Ahrens, 1948). In respect of double β-disintegration, mentioned above as a possibility for $^{130}_{52}Te$ (see also § 1.2, below), it might be pointed out that the availability of the coincidence method of detection in such cases increases the sensitivity of the direct search for possible radioactivity. Thus it has been concluded that the lifetime of $^{124}_{50}Sn$ against double β-decay is of the order 10^{17} years or longer (Fireman, 1949; Kalkstein and Libby, 1952)—representing a sensitivity at least three powers of ten better than could have been achieved by the usual method of single counting (see also Lawson, 1951).

1.2. Neighbouring stable isobars

The problem of the relative stability of isobaric nuclear species (that is, nuclear species characterized by a single value of the mass number, A) is closely linked with the problem of capture transformation which we have just considered. For that reason it may be discussed at this stage, although the stability rule involved was not one of the earliest to be enunciated when the results of the mass analysis of the elements first came under review. The rule is, briefly, that pairs of isobaric stable nuclei differ in charge number (Z) very much more frequently by two units than by one (Meitner, 1926). Current tables of 'existing' stable nuclei contain sixty-five examples† of the former relation and only two of the latter (it is assumed that the free neutron is β-active (Snell, Pleasanton and McCord, 1950; Robson, 1950, 1951). The two pairs of neighbouring isobars, for each of which both members are currently regarded as stable, are $^{113}_{48}Cd$ and $^{113}_{49}In$, and $^{123}_{51}Sb$ and $^{123}_{52}Te$. It is probably not without significance that these species lie so close together in the sequence of the elements.

The reason for the infrequent occurrence of neighbouring stable isobars is not far to seek. Let us consider the neutral atoms of two neighbouring isobars, $\begin{pmatrix} A \\ Z \end{pmatrix}$ and $\begin{pmatrix} A \\ Z+1 \end{pmatrix}$. If the masses of these two

† Eight of these pairs may further be grouped as four triads in each of which the three charge numbers stand in arithmetic progression with common difference two units. The species concerned are: $^{96}_{40}Zr$, $^{96}_{42}Mo$ and $^{96}_{44}Ru$; $^{124}_{50}Sn$, $^{124}_{52}Te$ and $^{124}_{54}Xe$; $^{130}_{52}Te$, $^{130}_{54}Xe$ and $^{130}_{56}Ba$; and $^{136}_{54}Xe$, $^{136}_{56}Ba$ and $^{136}_{58}Ce$.

atoms are not identical one or other atom must have the greater mass. Let us suppose, to begin with, that the neutral atom $\binom{A}{Z}$ is heavier than the neutral atom $\binom{A}{Z+1}$. Then, if it is assumed that no uncharged particle of finite rest-mass is emitted from the nucleus in the process of β-disintegration (that is, in contemporary theoretical phraseology, that the rest-mass of the neutrino is zero), and if the mass-difference between the two neutral atoms is greater than a small quantity $W_v(Z+1)/c^2$ which we shall presently define, it must be energetically possible for the species $\binom{A}{Z}$ to transform spontaneously into the species $\binom{A}{Z+1}$ by β-emission. In this case, clearly, both species cannot be stable.

Consider now the other alternative, namely that the mass of the neutral atom $\binom{A}{Z+1}$ is greater than the mass of the neutral atom $\binom{A}{Z}$. Then, making the same assumption concerning the mass of any hypothetical uncharged particle emitted in the process, if the difference in mass between the neutral atoms is greater than $W_K(Z)/c^2$, where $W_K(Z)$ is the K-ionization energy for an originally neutral atom of charge number Z, K-electron capture will be energetically possible for the species $\binom{A}{Z+1}$, and the assumption that the neighbouring isobars $\binom{A}{Z}$ and $\binom{A}{Z+1}$ are both essentially stable is obviously untenable. Again, in principle, at least (Marshak, 1942), a similar statement relates to transformation by capture of a less tightly bound extranuclear electron,† if the appropriate ionization energy, $W_L(Z)$, $W_M(Z)$, ..., is substituted for $W_K(Z)$.

We come to this conclusion, therefore (always assuming zero rest-mass for the neutrino, a result to which the experimental evidence steadily approximates (Hanna and Pontecorvo, 1949; Curran, Angus and Cockroft, 1949; Kofoed-Hansen, 1947, 1951)),

† L-electron capture has been experimentally detected for $^{37}_{18}A$ by Pontecorvo, Kirkwood and Hanna (1949).

that two neighbouring isobars cannot both be stable unless

$$-W_v(Z+1)/c^2 < M\binom{A}{Z+1} - M\binom{A}{Z} < W_v(Z)/c^2.\dagger$$

Here $W_v(Z)$ is the first ('valency') ionization energy and $M\binom{A}{Z}$ is the mass of the neutral atom $\binom{A}{Z}$, c being the velocity of light.

Since $W_v(Z)/c^2$ is of the order of 10^{-8} mass unit for all Z, the condition for the absolute stability of neighbouring isobars of any mass number is that the masses of the neutral atoms shall be the same within this narrow margin. Since there is no fundamental reason for such near identity of mass, the masses of neutron and proton differing by $1{\cdot}4\times10^{-3}$ mass unit, any case in which the condition is satisfied must be a case of coincidence. It is extremely unlikely, then, that there should even be two such cases in the system of the stable elements of natural occurrence, and we are led to suppose that one or other of the species in each of the pairs already listed is in fact unstable, though its disintegration is evidently of a highly forbidden type. It may be noted that, but for their recognized activity, the long-lived species $^{40}_{19}K$, $^{87}_{37}Rb$, $^{115}_{49}In$, $^{176}_{71}Lu$ and $^{187}_{75}Re$ would form neighbouring stable-stable pairs with $^{40}_{20}Ca$ (and $^{40}_{18}A$), $^{87}_{38}Sr$, $^{115}_{50}Sn$, $^{176}_{72}Hf$ (and $^{176}_{70}Yb$) and $^{187}_{76}Os$. In another section (p. 24) evidence will be presented which suggests that $^{113}_{48}Cd$ and $^{123}_{51}Sb$ are probably the unstable (β-active) members of the anomalous pairs.

At this point a somewhat academic remark is worth making, particularly because of its relevance for a later chapter (p. 118). It is that the conditions of stability of bare nuclei are different from those which have just been given for neutral atoms. Thus a little consideration will show that for the bare nucleus $\binom{A}{Z}$ β-disintegration to form the nucleus $\binom{A}{Z+1}$ is possible only when

$$M\binom{A}{Z} - M\binom{A}{Z+1} > \{W(Z+1) - W(Z)\}/c^2,$$

† It will be obvious that the limits would be further narrowed, by the replacement of $-W_v(Z+1)/c^2$ by zero, if β-disintegration into a bound (valency) state were regarded as possible (Daudel, Jean and Lecoin, 1947; Jean, 1948; Sherk, 1949; Ivanenko and Lebeder, 1950).

$W(Z)$ being the total energy necessary to ionize the atom completely—a quantity of the order of $5 \cdot 5 W_K(Z)$ (Allard, 1948; Foldy, 1951). Similarly, for positron emission to be possible from the bare nucleus $\binom{A}{Z+1}$, the masses of the neutral atoms must satisfy the inequality

$$M\binom{A}{Z+1} - M\binom{A}{Z} > 2m - \{W(Z+1) - W(Z)\}/c^2,$$

whereas, when the initial and final atoms are themselves neutral, the corresponding condition is

$$M\binom{A}{Z+1} - M\binom{A}{Z} > 2m,$$

m being the electronic mass.

Obviously, the distinction here made presupposes the reality of some kind of coupling between the energy states of the nucleus and of the outer atom; that supposition being granted we see that total ionization of an atom may, in the marginal case, suppress the radioactivity of a species which is normally β-active† or bring to light the latent positron activity of a species which appears positron-stable in the neutral state. It is hardly necessary to point out that the limits for the incidence of this effect are narrow ones, being of the order of $10^{-7} Z$ mass unit, or about $150Z$ eV., but it seems that at least one example of its operation has been recognized.‡ As regards the capture process, there is no sense, of course, in speaking of K-electron capture by a bare nucleus, but it will appear from what has previously been said that this type of transformation is energetically most probable when the initial atom is already completely ionized except for a single electron in the K shell—and a similar remark holds for L-, M-, ... electron capture, as the case may be. Ultimately, capture of a free electron by a bare nucleus is seen as energetically most favoured of all.

Having established a satisfactory explanation of the rarity of pairs of neighbouring 'stable' isobars, it is interesting to return to

† As before we are neglecting β-disintegration into bound states. Such disintegration cannot similarly be suppressed by ionization of the atom.

‡ Indications are that the energies of β-disintegration for the predominant modes of $^{227}_{89}Ac$ and $^{228}_{88}MsTh_1$ are less than $150Z$ eV (~ 13 keV) in each case.

the question of the frequency of occurrence of pairs of apparently stable isobaric species for which the difference of charge number is two. For it will be recognized that the type of analysis already given is equally valid for this case, if double β-disintegration and double electron capture are accepted as possible unit events. On this basis, then, one or other member of each such pair of isobars must in the last resort be regarded as essentially unstable. The fact that no example of a double process—either double β-emission or double electron capture—has with certainty been established (experimental limits of stability have already been given for two species, see p. 5)† enforces the conclusion that the disintegration constant for such a process is always very small (say, $< 10^{-23}$ sec.$^{-1}$) even when the available energy is considerable (say, several MeV.). This conclusion can be reconciled with current theory (Touschek, 1948; Daudel and Jean, 1949) without much difficulty.

One other conclusion can clearly be drawn from the experimental facts. It is that if $\binom{A}{Z}$ and $\binom{A}{Z+2}$ are two 'stable' isobars differing in charge number by two, then the 'unobserved' intermediate species $\binom{A}{Z+1}$ must be essentially unstable in respect both of β-disintegration and of electron capture. If it were not unstable in respect of either transformation, then obviously the rule regarding neighbouring isobars would be violated for one pairing or the other, that is, either for the pairing $\binom{A}{Z}$, $\binom{A}{Z+1}$ or for the pairing $\binom{A}{Z+1}$, $\binom{A}{Z+2}$. In fact the predicted branching disintegration of these unstable intermediate species has so far escaped detection in the majority of cases (Feather, 1948*a*): clearly in each of these cases one disintegration mode is much more probable than the other.

1.3. The distinction between even and odd

The sixty-one‡ values of A, for which pairs of stable isobars differing by two units in Z are known to exist, are all even values—and the values of Z belonging to these paired species are also,

† Levine, Ghiorso and Seaborg (1950) have recently shown that the half-value period for the process ${}^{238}_{92}U \xrightarrow{2\beta} {}^{238}_{94}Pu$ is greater than 6×10^{18} years.

‡ See footnote, p. 6.

without exception, even. This is a very striking regularity, and it is in line with many other empirical generalizations, all of which tend to emphasize the much greater frequency of occurrence of even numbers than odd as descriptive of nuclear structure. It is an aspect of this regularity that the 'unobserved' intermediate species discussed at the end of the last section are all of odd charge number and even mass number—and this fact itself is contained in a wider generalization, already well known (Russell, 1923; Harkins, 1933), that for $Z > 7$ no species† of the same type (A even, Z odd)‡ has been observed as a stable species in nature. We shall come upon a further aspect of the same generalization when we discuss regularities in the disintegration energies of β-active bodies as a whole (p. 139).

The generalization that for $Z > 7$ species of the type $\binom{\eta}{\omega}$ are essentially unstable is reinforced by the further empirical rule that, within the same range of odd values of Z, elements are either simple, that is, are represented by a single stable $\binom{\omega}{\omega}$ species only, or else they have two stable isotopes of type $\binom{\omega}{\omega}$ separated in mass number by two units. There is no known exception to this rule.§ It may be formulated, on the basis of the currently accepted neutron-proton model of the nucleus, by the statement that when the number of protons in any stable nucleus is odd (and greater than 7) the number of neutrons is always even, it being recognized that stability cannot be attained, for the same (odd) value of Z, for more than two different (consecutive) even values of the neutron number, N.‖ The tendency towards the pairing of neutrons—and, as we shall see, also of protons—in the nucleus, which this formulation emphasizes, appears to be a universal tendency when fundamental particles of

† With the possible exception of $^{50}_{23}V$, the radioactivity of which has yet to be detected.

‡ Where convenient we shall use the abbreviations $\binom{\eta}{\eta}$, $\binom{\eta}{\omega}$, etc., for the nuclear types which in respect of mass and charge numbers $\binom{A}{Z}$ are $\binom{\text{even}}{\text{even}}$, $\binom{\text{even}}{\text{odd}}$, etc.

§ Again, unless $^{50}_{23}V$ provides such an exception.

‖ When we wish to represent nuclear types in relation to proton and neutron numbers we shall use the abbreviations $\binom{e}{e}$, $\binom{e}{o}$, etc., for $\binom{N}{Z}$ $\binom{\text{even}}{\text{even}}$, $\binom{\text{even}}{\text{odd}}$, etc.

'half-integral' spin are involved. Thus, the empirical fact that all nuclei, except the very lightest,† which contain odd numbers both of protons and of neutrons, are either β-active or capture-active (or both) may be restated broadly as follows: when the numbers of protons and neutrons are both odd, either the transformation of the unpaired neutron and unpaired proton into a pair of protons, or their transformation into a pair of neutrons (or in some cases either transformation), will result in a system of greater stability. It will be noted that the effect of any such transformation is to restore both the neutron and proton numbers of the nucleus to even values.

Turning now to species of even Z we come upon generalizations which have a similar basis of explanation. For even Z, in the range $7 < Z < 83$, there is no element with less than three stable isotopes, and one ($_{50}$Sn) has ten. The average, for the 38 elements of even Z included in this range, is 5·7 stable isotopes each. Without exception, the stable isotopes both of lowest and of highest mass number, for each of the 38 elements in question, are of even A. On the basis of this result alone it is evident that there are more species of even than of odd A belonging to these elements of even Z, but the disparity is greater than would follow from this result alone. Of 216 stable species belonging to the 38 elements, 53 are of odd A and 163 of even A. With the single exception of tin (which has odd-A isotopes $^{115}_{50}$Sn, $^{117}_{50}$Sn and $^{119}_{50}$Sn), the elements of even Z follow the same rule as do those of odd Z in respect of their stable isotopes of odd A. With the exception noted, this rule can thus be stated generally: no element possesses more than two stable isotopes of odd A, and when it possesses two such isotopes the mass numbers involved are consecutive odd numbers. To extend the rule in this way, however, tends to obscure a very real point of comparison. The significant comparison would appear to be between regularities relating to nuclei containing a given odd number of protons and an even number of neutrons and those containing a specified odd number of neutrons and an even number of protons, that is between stable nuclei of the types $\begin{pmatrix}\omega\\\omega\end{pmatrix}$ or $\begin{pmatrix}e\\o\end{pmatrix}$, Z constant, and $\begin{pmatrix}\omega\\\eta\end{pmatrix}$ or $\begin{pmatrix}o\\e\end{pmatrix}$, N constant, of our earlier classifications.

† That is, except the $\begin{pmatrix}\eta\\\omega\end{pmatrix}$ nuclei $^2_1$H, $^6_3$Li, $^{10}_5$B and $^{14}_7$N.

If we make a survey of the $\binom{\omega}{\eta}$ species from this point of view we find an interesting result. There are only three examples, in the current lists of existing species, of pairs of stable nuclei having the same odd number of neutrons $(N > 7)$; apart from these cases, when the number of neutrons is odd it seems that stability is attained only for a single (even) number of protons. Moreover, the three 'exceptional' cases all carry an element of doubt. It would appear that the true position might almost be stated: when the number of nuclear protons is odd $(Z > 7)$ stability may be attained for a single even number, or for two (consecutive) even numbers, of neutrons; when, however, the number of nuclear neutrons is odd $(N > 7)$ stability is attained only for a single (even) number of protons. This is one clear aspect of an essential asymmetry as between neutrons and protons as structural units of nuclei. We shall encounter further evidence of this asymmetry at a later stage (pp. 66, 80, 123, 136).

The three 'exceptional' pairs just mentioned are $^{97}_{42}$Mo and $^{99}_{44}$Ru, each containing 55 neutrons, $^{113}_{48}$Cd and $^{115}_{50}$Sn, with 65 neutrons, and $^{145}_{60}$Nd and $^{147}_{62}$Sm, with 85 neutrons each. The doubt in the first case is connected with the question of the 'missing' element Tc(43) which will be considered in the next section (p. 16), in the second case the exception would disappear if $^{113}_{48}$Cd were to prove to be unstable—as we have already suggested is likely, and the third pair includes $^{145}_{60}$Nd which is possibly β-active (pp. 4, 17). It is evident, then, that these three $\binom{\omega}{\eta}$ pairs are correctly described as exceptional.

The class of nuclei which we have not so far discussed systematically is that for which both charge and mass numbers are even. These $\binom{\eta}{\eta}$ nuclei obviously contain even numbers both of protons and neutrons, and they are, as a class, the most stable and of by far the most frequent occurrence on the earth. As we have already noted, the lightest and heaviest of the stable isotopes of elements of even $Z (Z > 7)$ are $\binom{\eta}{\eta}$ species, and, in the same series of elements, the isotope of greatest terrestrial abundance is also an $\binom{\eta}{\eta}$ species in all cases—with the single exception of platinum, for which the $\binom{\omega}{\eta}$

species $^{195}_{78}$Pt is the most abundant. Concerning nuclear spin, there is nothing to contradict the general impression that the spin is zero for all $\binom{\eta}{\eta}$ nuclei which are stable. This provides further evidence that the pairing of neutrons and protons—separately—is an anti-parallel spin pairing.

With the lightest elements there is a particular class of $\binom{\eta}{\eta}$ species which calls for separate mention. This is the class for which $N=Z$ and for which, therefore, the mass number, A, is a multiple of four. The α-particle, $^4_2$He, is the simplest of these nuclei. As is clear from a consideration of 'exact' masses, it is by far the most tightly bound of the nuclei with $A<8$. The other nuclei in the class which we are considering may, formally at least, be regarded as aggregates of α-particles. Extending as far as $^{40}_{20}$Ca, this sequence of nuclei, with two exceptions, $^8_4$Be and $^{36}_{18}$A, contains the predominant isotope of each of the elements of even $Z(Z<21)$. Including $^{12}_6$C, $^{16}_8$O, $^{24}_{12}$Mg, $^{28}_{14}$Si, as well as $^{40}_{20}$Ca, it contains most of the nuclear species of really common occurrence on the earth. As regards the exceptions, that involving $^{36}_{18}$A is probably less real than appears. It is more than likely that the excessive abundance of $^{40}_{18}$A in atmospheric argon is connected with the presence in the lithosphere of the unstable, though long-lived, $^{40}_{19}$K (v. Weizsäcker, 1937; Aldrich and Nier, 1948; Inghram, Brown, Patterson and Hess, 1950), practically the whole of the observed $^{40}_{18}$A having been produced by the capture-decay of this species during terrestrial times. If this is so, it may still be correct to think of $^{36}_{18}$A as the most abundant 'natural' isotope of argon. As regards $^8_4$Be, the fact that very large amounts of energy (~ 16 MeV.) are liberated in the β-disintegrations of $^8_3$Li and $^8_5$B shows that $^8_4$Be is by far the most nearly stable nucleus of mass number 8: but for its α-activity† it would probably be more abundant than $^9_4$Be. In view of this consideration its non-occurrence as a truly stable species appears less exceptional than might otherwise be concluded. This non-occurrence provides a reason, however, for not attaching undue significance to the formal description of the '$A=4n$' nuclei as aggregates of α-particles—for adopting this

† Hemmendinger, 1948, 1949; Tollestrup, Fowler and Lauritsen, 1949; Carlson, 1951.

view it would appear rather difficult to understand why a system composed of three or four α-particles should be stable when one composed of two α-particles is unstable (Massey and Mohr, 1936; Wheeler, 1937, 1941; Haefner, 1951).

Whilst discussing the question of the 'predominant' isotopes of the lighter elements of even Z (and, in those cases already considered, 'predominant' in fact refers to an isotopic abundance always greater than 77 %) it is worth while to extend our survey slightly beyond $Z=20$. In so doing we find one isotope almost equally predominant for each of the next three elements of even Z, but in these three cases ($^{48}_{22}Ti$, $^{52}_{24}Cr$ and $^{56}_{26}Fe$) we have $N-Z=4$, rather than $N=Z$. This is the beginning of a regularity to which we shall later devote a separate section, when the significance of the number $N-Z$ (the excess of the number of neutrons over the number of protons in the nucleus) as a parameter for use in the classification of nuclei will be further discussed (p. 19).

1.4. 'Missing' stable species

So long as evidence regarding the stability or instability of nuclear species is based entirely on the results of mass-spectrum analysis and the investigation of naturally occurring elements for weak radioactivity, the situation in relation to many 'unobserved' species of possible stability is open to considerable doubt. It is true that during recent years great improvements in the sensitivity of mass analysis have been made—a sensitivity of a few parts in a million of the total element being reached in favourable cases (Duckworth, Black and Woodcock, 1949; Leland, 1949 *a*)—but, in general, a much higher degree of certainty (regarding instability) derives from observations on artificially produced radioactivity. Moreover, in point of time, a great deal of information on this topic was available before the significant improvements in methods of mass analysis were brought about. The first artificially produced radioactive isotope was reported in 1934; the improvements in the mass spectrograph date from about ten years later. At the best, mass analysis of naturally occurring substances may add a few species per year to the list of observed, and therefore presumably stable, species—and the return will obviously be a rapidly diminishing one—on the other hand, observations on 'artificial radioactivity', over a period of fifteen years, have established without question the essential instability of some 500 species ($Z < 84$), making it impossible for us to believe that these species will ever appear in the mass spectrum of a terrestrial element of normal provenance, whatever improvement in instrumental performance is ultimately achieved. These are general remarks: their relevance should become more apparent as we pass to the systematic consideration of the problem of the missing (unobserved) species. In this connexion we consider first missing values of Z, and then missing values of N, in the table of stable nuclei.

It has been known for many years that there are just two missing values of Z in the range $0 < Z < 84$, which is otherwise filled with the atomic numbers of known elements—either simple elements, or elements possessing two or more stable isotopes. This was the situation in 1934, before the discovery of 'artificial radioactivity', and since that time no greater success has attended any attempt to detect either of these missing elements in terrestrial minerals. The atomic numbers of the missing elements are 43 and 61.

Let us consider first what the mass numbers of the possible stable isotopes of these elements might be. Because we are dealing with elements of odd Z the possibilities are strictly limited. A study of the existing species of near-neighbouring Z (cf. § 1.5) indicates that for $Z=43$ there are only two such possibilities, $A=97$ and $A=99$, whilst for $Z=61$, mass numbers 145, 147 or 149 could conceivably belong to stable isotopes. Now current tables already contain $^{97}_{42}$Mo and $^{99}_{44}$Ru as stable, so it might appear that neither of the 'neighbouring' isobars with $Z=43$ could be allowed. However, it must be remembered that we have already found reason to regard the pair $^{97}_{42}$Mo and $^{99}_{44}$Ru as in one sense exceptional (p. 13). The position, therefore, must be examined further. A very similar situation obtains in respect of the possible stable isotopes of $Z=61$. For each a neighbouring 'stable' isobar exists (namely $^{145}_{60}$Nd, $^{147}_{62}$Sm† and $^{149}_{62}$Sm, corresponding to the three values of A listed above), but again two of these ($^{145}_{60}$Nd and $^{147}_{62}$Sm) form a pair analogous to the pair, $^{97}_{42}$Mo, $^{99}_{44}$Ru, already under suspicion.

In essence this is the situation when the evidence from mass analysis alone is considered. It is clarified immediately by the

† The α-active $^{147}_{62}$Sm is assumed to be β- and capture-stable.

appeal to present knowledge of artificially produced radioactivity. This appeal almost settles the question so far as the missing element 43 is concerned, and goes a long way towards settling it for $Z=61$. For $Z=43$, the species of mass number 97 is known in its isomeric state of half-value period 93 ± 5 days, the assignment having been made by studying the activities obtained by bombarding the separated isotopes of molybdenum with deuterons (Motta and Boyd, 1948), and in its ground state is probably capture-active and of long life, and the species of mass number 99 is a long-lived β-emitter of lifetime a few hundred thousand years, identified directly by mass analysis (Inghram, Hess and Hayden, 1947). For $Z=61$, $A=147$ is known to belong to a β-emitter of half-value period about 3 years, and $A=149$ to another β-emitter of period 54 hr. Identification of the former of these species has the certainty of mass analysis (Hayden, 1948), that of the latter the strong support of evidence regarding the generic relations of production and decay (Marinsky, Glendenin and Coryell, 1947). For the two elements, only the possibly stable species $^{145}_{61}$Pm remains unidentified (Ballou, 1948) as, in fact, one of the 'artificial' products of fission or other induced nuclear change. This species is certainly β-stable, for $^{145}_{62}$Sm is capture-active (Inghram, Hayden and Hess, 1947*a*); if it is capture-stable, also, then $^{145}_{60}$Nd must be β-active.† If this is so, then in all probability $^{145}_{61}$Pm is itself α-active, otherwise there would be no good reason why it should not occur as a stable species of normal abundance on earth. In upshot, element 43 appears to be a truly 'missing' element in that it is known that it has no possible stable isotope, and element 61 is equally truly 'missing'—unless the non-appearance of $^{145}_{61}$Pm is due solely to its α-activity. Furthermore, the species, $^{97}_{42}$Mo, $^{99}_{44}$Ru, remain as infringing the general rule that for N odd (and greater than 7) there is only one value of Z (for each N) corresponding to stability—and $^{145}_{60}$Nd and $^{147}_{62}$Sm as a second example of such infringement, if in fact $^{145}_{61}$Pm is capture-active. As regards the third example of this infringement ($^{113}_{48}$Cd and $^{115}_{50}$Sn), it may here be noted that if this infringement were to be definitely established, and if the neighbouring-isobar rule were

† Since the above, and other remarks (pp. 13, 23, 72) on the problem of the stability of $^{145}_{60}$Nd and $^{145}_{61}$Pm were written, a report by Butement (1951) appears to show that $^{145}_{61}$Pm is, in fact, capture-active. Confirmation of this result will be awaited with interest (see Mulholland and Kohman, 1952).

to be taken as absolute, then indium ($Z=49$) should be added to the list of missing elements. In this connexion Kowarski (1950) has recently emphasized the total absence of stable isotopes of odd A for $Z=18$ (argon) and $Z=58$ (cerium). In this respect, at least, these two even-numbered elements also are 'missing' in nature.

We have now discussed the problem of missing stable species from the point of view of charge number Z, that is, from the point of view of the number of protons in the nucleus, and we have found that the missing species are represented by the odd numbers 43 (certainly) and 61 (possibly), noting that further close investigation may conceivably add the odd number 49 to this list. We should here recall the basic assumption underlying our discussion; it is the assumption that the whole structure of the system of the stable elements in the range $Z<84$ is determined by questions of stability to β- (or capture-) transformation—species which are merely α-active may, for our main purpose, be regarded as stable. We have next to discuss the problem of missing species from the point of view of the neutron number N.

Instead of some two missing species, when classification proceeds according to Z values, we find nine missing species of nuclei in a classification according to N. No stable nuclei are known which contain the following numbers of neutrons, namely, 19, 21, 35, 39, 45, 61, 89, 115 and 123. It will be noted that all these missing species are characterized by odd neutron numbers; this, of course, is in part a reflexion of the fact that stable nuclei of the type $\binom{\eta}{\omega}$ or $\binom{o}{o}$ do not occur for $Z>7$. On the other hand, the present aspect of the matter emphasizes again the feature of asymmetry which we have already recognized as between the status of protons and neutrons in the nucleus. For, as concerns the $\binom{\omega}{\eta}$ species which might occur with the nine odd values of N listed above, it can be stated at once that 18 of the 19 possible† species in this group must be unstable if the rule regarding neighbouring isobars is accepted as absolute. And the nineteenth of these species, $^{149}_{60}Nd$, is known to be unstable on the evidence of artificial radioactivity alone. It would therefore

† That is species having (odd) mass numbers within the limits of the greatest and least (even) mass numbers characterizing the known stable isotopes of the (even-numbered) elements concerned.

appear that, from the point of view of classification according to neutron number, there are in fact nine 'missing' neutron-number types in the range $N < 127$. Nevertheless, it is difficult to recognize any suggestive regularity in the sequence of N values concerned, and the detailed implications of our analysis remain in consequence obscure. In particular it may be no more than a coincidence that the number 61 occurs (provisionally) both in the list of missing Z and in that of missing N.

1.5. The significance of the isotopic number, $A-2Z$

Reference has already been made to predictions based upon surveys of existing stable species, within certain restricted ranges of Z, when the possibilities in respect of stability of unknown types were under discussion. In order effectively to make a survey of this type implies, of course, that some sort of scheme of arrangement is used, whereby such regularities as exist are given due prominence. Several schemes of this kind have been developed in the past (Fournier, 1929; Turner, 1940; Kohman, 1948), but we shall confine our attention to one of them at this stage. This is the scheme of presentation which employs, along with the mass number A, or the charge number Z, the parameter $A-2Z$, to which the name isotopic number has been given (Harkins, 1921 *a*). It will be obvious that this parameter is $N-Z$, the excess of the number of neutrons over the number of protons in any nucleus, and the name isotopic number will be self-explanatory, in that the sequence of values of $A-2Z$ for the isotopes of any element corresponds exactly with the sequence of the mass numbers of the isotopes in question.

The choice as to whether the isotopic number should be used together with the mass number or with the charge number for purposes of presentation is not very critical, but the usage of fig. 1, in which A is plotted as ordinate and $A-2Z$ as abscissa, will be followed in this monograph. Such an arrangement represents α-disintegrations by vertical and β-disintegrations by horizontal displacements of the representative point. This has the merit of clarity, particularly as we have agreed that the configuration of points representing the stable nuclei is almost entirely determined by questions of β- (or capture-) stability. Stability limits are indicated, therefore, by the horizontal width of the domain of points representing the stable nuclei in the figure. We shall be discussing the trend of this domain in more detail at a later stage (p. 70); meanwhile it may be noted that it has two well-marked constrictions, at about $A=88$ or 90 and $A=140$ or 142. It is possibly significant that these recur at roughly equal intervals of 50 in A beyond $A=40$, up to which point the 'left-hand' limit of stability (i.e. the limit of capture-stability) is given by the line $A-2Z=0$.

We can now, with profit, re-formulate some of the generalizations already discussed in terms of their representation on the $A/(A-2Z)$ diagram for stable nuclei. Since the isotopic number is even or odd as A is even or odd, the rule that no stable nuclei of the type $\binom{\eta}{\omega}$ exist for $Z>7$ reduces to the statement that for values of $A-2Z$ which are even multiples of two, only those values of $A (A>14)$ which are also even multiples of two, (i.e. values of A of the type $4n$, with n integral) are possible—and that for values of $A-2Z$ which are odd multiples of two only A values $(A>14)$ which are odd multiples of two can occur. Whatever the (even) value of the isotopic number, therefore, possible A values for the stable nuclei must be members of a series of common difference four units. That this is so—and that in general for a given (even) value of $A-2Z$ the series of 'stable' A values is an unbroken series (i.e. without missing members, between specific limits in each case)—is evident from fig. 1. A study of fig. 1 shows that, for odd values of the isotopic number, also, obvious series of 'stable' A values occur—again, in general, unbroken series—though the common difference is now two units in A in each case. This distinction between the series for odd and even values of the isotopic number reflects the fact that odd values of A, and thus of $A-2Z$, can occur both with odd and with even values of Z, whereas even values of A belong exclusively to even values of $Z (Z>7)$.

Having recognized these broad generalizations in their new form, it is interesting to examine the small number of cases in which irregularities appear in the series which the $A/(A-2Z)$ diagram reveals—for the occurrence of generally unbroken series, each having $A-2Z$ constant, is a positive feature of the system of the stable species (Harkins, 1921*b*; Russell, 1923, 1924*a*, 1924*b*) which is not implicit in the original generalizations, these being essentially

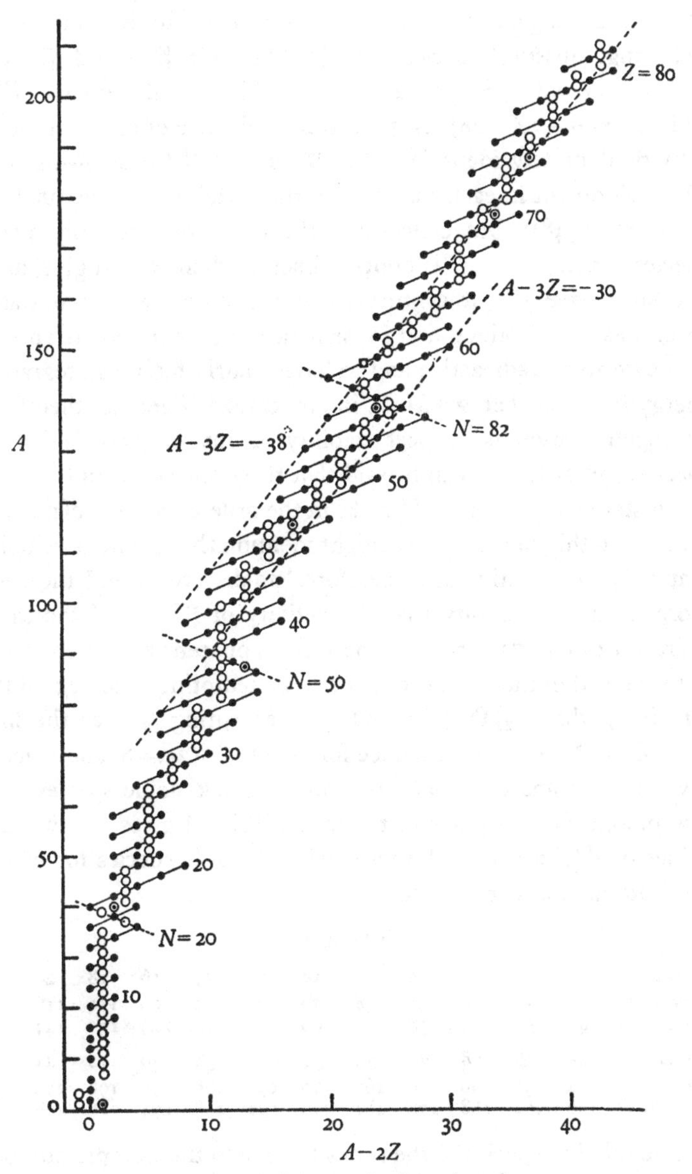

Fig. 1. 'Naturally occurring' species, $A<210$.
○ odd A, ● even A, ⊙ β-active, ◘ α-active.

restrictive in character. So far as even values of isotopic number are concerned the gaps in the otherwise unbroken series indicate the following 'missing' species ($Z > 7$): ${}^{52}_{22}$Ti, ${}^{56}_{24}$Cr, ${}^{60}_{26}$Fe and ${}^{88}_{40}$Zr (all with $A - 2Z = 8$), ${}^{90}_{38}$Sr ($A - 2Z = 14$), ${}^{140}_{60}$Nd ($A - 2Z = 20$) and ${}^{140}_{56}$Ba and ${}^{144}_{58}$Ce ($A - 2Z = 28$). Appeal to the evidence of artificial radioactivity definitely identifies ${}^{90}_{38}$Sr, ${}^{140}_{56}$Ba and ${}^{144}_{58}$Ce as β-unstable (the assignment in each case having the certainty of mass analysis (Hayden, 1948)), but concerning the other five possibly stable species nothing definite is known. Each of them would give, after two successive β- (or capture-) disintegrations, a known stable species as end-product, and in each case the intermediate species is known to be radioactive and to have a fairly high disintegration energy (not less than 2·2 MeV. in any case).† There is, therefore, no cogent reason to suspect stability of any of these 'missing' species, but at least it can be said that these species, if in fact they are unstable, represent real breaks in the ordered system of nuclear species. In this connexion we might mention the remote possibility that ${}^{140}_{60}$Nd is α-active and therefore 'unobserved', and the even more distant possibility that the activity of ${}^{88}_{40}$Zr is of the same character (see p. 73). Even if these two apparent exceptions were not removed in this way, it might still be tempting to disregard the remaining three, ${}^{52}_{22}$Ti, ${}^{56}_{24}$Cr and ${}^{60}_{26}$Fe, accepting ${}^{64}_{28}$Ni as the first normal member of the sequence for which $A - 2Z = 8$, and placing the whole anomaly in this series on the freak stable species ${}^{48}_{20}$Ca (see p. 25). For purposes of reference, Table I gives the limiting values of A ($A > 14$) for all those series of stable species for which the isotopic number is even.

TABLE I

$A-2Z$	0	2	4	6	8	10	12	14	16	18	20	22
Lowest A	16	18	36	46	48	70	76	82	96	110	116	122
Highest A	40	58	64	78	96	106	112	114	124	130	144	142

$A-2Z$	24	26	28	30	32	34	36	38	40	42	44
Lowest A	124	130	136	150	160	170	176	186	192	198	204
Highest A	156	162	168	174	184	190	196	198	204	206	208

The missing species of the series for which the isotopic number is odd are slightly more numerous than those of the even series,

† As an example, ${}^{56}_{24}$Cr would give ${}^{56}_{26}$Fe by two β-disintegrations. Here the intermediate species is ${}^{56}_{25}$Mn, and its disintegration energy is known to be 3·6 MeV.

though, as we shall see, their number may be reduced from eleven to seven (or even four) mainly by a plausible interpretation of the singularity which we have already noted in the $A/(A-2Z)$ diagram in the neighbourhood of $A = 140$. Before admitting this interpretation we should be inclined to conclude that the following species are missing, namely: ${}^{37}_{18}A$, ${}^{39}_{18}A$, ${}^{97}_{43}Tc$, ${}^{99}_{43}Tc$, ${}^{115}_{49}In$, ${}^{137}_{57}La$, ${}^{139}_{58}Ce$, ${}^{145}_{61}Pm$, or ${}^{145}_{60}Nd$, ${}^{141}_{58}Ce$, ${}^{143}_{59}Pr$ and ${}^{147}_{61}Pm$. Many of these species we have already considered in other connexions; here the important fact is that for each there exists a stable neighbour. We conclude, therefore, that all are truly unstable—and we have to inquire what the significance of these many gaps in the series of odd isotopic number might be. This inquiry leads in the first place to the 'plausible interpretation' mentioned above. We notice that six of the missing species in the above list belong to the neighbouring odd values of isotopic number 23 and 25. A study of fig. 1 in the region of $A = 140$ gives the clue to the interpretation. The whole structure of the diagram in this region suggests that the sequence of series of odd isotopic number suffers a discontinuity here. It is as if ${}^{139}_{57}La$ is the last member of one sequence and ${}^{141}_{59}Pr$ the first member of a new sequence of such series, and as if the nature of the discontinuity derives from the ability of nuclei of mass number greater than 140 to accommodate a larger number of protons† (for a given total mass number) than would be expected on the basis of series regularities obtaining up to that point ($A < 140$). We shall be bringing forward other reasons later for supposing that a discontinuity of this type exists at this stage in the general process of the building-up of heavier from lighter nuclei (p. 70); for the present we shall accept the interpretation which has been given and incorporate it in Table II. This table shows the limiting values of A for the series of stable species of odd isotopic number, arranged under the headings of 'first sequence' and 'second sequence', respectively.

To accept the distinction between the two main sequences of series in Table II clearly makes the instability of the species ${}^{137}_{57}La$, ${}^{139}_{58}Ce$, ${}^{141}_{58}Ce$ and ${}^{143}_{59}Pr$ a matter of regularity; and the regularity would be further extended if ${}^{145}_{60}Nd$ were β-active and ${}^{145}_{61}Pm$ were stable, for then ${}^{147}_{61}Pm$ would also be 'regular'. In fact none of these

† At least two extra protons, as the form of the diagram clearly shows.

six† unstable species would any longer be 'missing' species according to our conventional definition, if these assumptions were made. Similarly, $^{115}_{49}$In would disappear from the list of exceptions if $^{113}_{48}$Cd were unstable rather than $^{113}_{49}$In‡ (if the reverse were true $^{113}_{49}$In would become an additional missing species according to our definition, see p. 18). Admitting all our assumptions, then, four certain irregularities would remain; they concern the instability of the four species $^{37}_{18}$A, $^{39}_{18}$A, $^{97}_{43}$Tc and $^{99}_{43}$Tc. It can hardly be fortuitous that all these species have mass numbers close to the values which are associated with two of the breaks in the $A/(A-2Z)$ diagram already distinguished (p. 20). In the separation of the two sequences in Table II, we have made an attempt to interpret the third of these breaks in terms of a real discontinuity in nuclear constitution; it is probable that, as more information is brought to bear on the problem, interpretation of all three will be possible along similar lines.

TABLE II

(*a*) First sequence

$A-2Z$	1	3	5	7	9	11	13	15	17	19	21	23	25
Lowest A	7	37	49	65	71	81	97	109	113	119	123	131	137
Highest A	39	47	63	69	79	99	107	115	117	123	129	135	139

(*b*) Second sequence

$A-2Z$	23	25	27	29	31	33	35	37	39	41	43
Lowest A	141	(145)	153	157	163	173	179	189	193	201	205
Highest A	147	151	155	161	171	177	187	191	199	203	209

Before leaving the question of the series of stable species of odd isotopic number, one further point may be remarked. It is that the naturally occurring β-active isotope of rubidium, $^{87}_{37}$Rb, has $A-2Z=13$. Comparison with Table II shows how far this species is from the normal range of stability for this value of the isotopic number; it might confidently have been predicted as being the unstable isotope of rubidium on the basis of these series regularities alone.

Above, we have effectively based predictions regarding instability

† Throughout this discussion $^{145}_{61}$Pm and $^{145}_{60}$Nd are regarded as alternative missing species (see p. 17).

‡ An isomeric state of $^{113}_{48}$Cd is known which is β-active (Carss, Gum and Pool, 1950; Cassidy, 1951), but it is at present uncertain whether the β-disintegration energy is greater than the excitation energy or vice versa.

in the pairs $^{145}_{60}$Nd, $^{145}_{61}$Pm and $^{113}_{48}$Cd, $^{113}_{49}$In on series regularities; in the last doubtful case, that of the pair $^{123}_{51}$Sb, $^{123}_{52}$Te (p. 8), the indications are less clear-cut. Neither member of the pair, being proved unstable, would thereby become a 'missing' species according to our definition. But the sequence of series of stable species of odd isotopic number from $^{119}_{50}$Sn to $^{139}_{57}$La would appear somewhat more uniform if, in fact, $^{123}_{51}$Sb were β-active than it would if $^{123}_{52}$Te were capture-active. We shall, therefore, leave prediction at this point—all too conscious of its hazards.

1.6. Favoured values of Z and N

In the last two sections attention has been concentrated upon the gaps in the ordered 'series' of species which any systematic study of the existing stable nuclei reveals. We now turn from the consideration of these 'missing' species to the complementary problem of the occurrence of 'unexpected' or 'supernumary' species, that is to the problem of the concentration of existing species in 'favoured' elements (favoured proton-number types) or in favoured neutron-number types. The facts that there is clear evidence for such concentration, and that features of regularity emerge from its consideration, provide additional reasons for the belief that the breaks in order are significant.

From what has already been said we must expect that the favoured proton and neutron numbers are even numbers. In respect of proton numbers the even numbers concerned are the group around 50, and the number 20. The only element possessing ten stable isotopes is tin ($Z=50$), xenon ($Z=54$) has nine, and cadmium ($Z=48$) and tellurium ($Z=52$) have eight each. No other element has more than seven stable isotopes. Calcium ($Z=20$) is the only element with atomic number less than 36 for which the number of stable isotopes is as great as six. So far as these elements are concerned $^{136}_{54}$Xe and $^{48}_{20}$Ca (see Jones and Kohman, 1952) are 'unexpected' or 'supernumary' species (the complementarity of present considerations with those of the last section will be clearly evident from this statement).

Favoured neutron numbers are somewhat more numerous than favoured values of Z, and they extend over a somewhat wider range. The position is that each of forty-six of the sixty even neutron-number types having $8 \leqslant N \leqslant 126$ is represented either by three or by four stable species—and six other such types have only two stable representatives each: of the remaining eight neutron-number types, the seven for which $N=20$, 28, 50,† 58, 74, 78 and 90 each have five representatives, and the eighth ($N=82$) has seven.‡ We regard these eight neutron numbers, 20, 28, 50, 58, 74, 78, 82 and 90, then, as the favoured numbers. It will be observed that this list again gives prominence to the numbers 20 and 50, and in addition clear indication of the significance of added multiples of 8—particularly taking $N=50$ as base. In this last respect $N=78$ is strictly out of place and $N=66$ is missing (for $N=66$ to be included $^{118}_{52}$Te and $^{112}_{46}$Pd should be stable: the position is that $^{118}_{52}$Te is at present not known with certainty (Lindner and Perlman, 1948; Leland, 1949 *a*), but $^{112}_{46}$Pd is β-active (Nishina, Yasaki, Kimura and Ikawa, 1940; Segrè and Seaborg, 1941)), but the reality of the periodicity is clear enough. Furthermore, 82, which is the most-favoured of the neutron numbers, is also the greatest even proton number for which ordinarily stable species exist. Above this value of Z natural α-activity sets in.

We are compelled, therefore, to the final conclusion that 8, 20, 50 and 82 are the primary ('magic') numbers (Harkins, 1949) for any theory of the structure of the nucleus (see Mayer, 1948)—and the temptation is great to embark at once on an ambitious attempt to find their interpretation in a 'shell model', somewhat after the pattern of the Bohr model of atomic electron shells,§ or in a static geometrical model deriving ultimately from ideas of crystal structure (Rutherford, 1924; Wefelmeier, 1937; Winans, 1947). That temptation we shall, for the present, resist.

† A sixth species with $N=50$ is the long-lived $^{87}_{37}$Rb, the anomalous position of which in fig. 1 has already been noted (p. 24).

‡ This last group carries $^{144}_{62}$Sm, a third unexpected species.

§ Bartlett, 1932; Jones, 1932; Gapon, 1932, 1933; Landé, 1933 *a*, 1933 *b*; Elsasser, 1933, 1934 *a*, 1934 *b*; Guggenheimer, 1934 *a*, 1934 *b*.

CHAPTER II

REGULARITIES IN α-DISINTEGRATION

2.1. The Geiger-Nuttall rule

The Geiger-Nuttall rule is the expression of the empirical fact that there exists a fairly obvious, approximately monotonic, relation between disintegration constant and particle energy for the α-active bodies of each of the three 'classical' series of radioelements (Geiger and Nuttall, 1911, 1912*a*, 1912*b*). As originally exhibited, this relation, for any series, was conceived in terms of the logarithms of the disintegration constants (in sec.$^{-1}$) of the various α-active bodies belonging to the series, and the logarithms of the ranges (in cm. in 'standard' air) of the α-particles emitted by these bodies. It appeared that the relation was approximately linear in each case and, in fact, that the slopes of the three Geiger-Nuttall lines were very closely the same, so that the result could be expressed formally by the equation

$$\log \lambda = a + b \log R, \qquad (2.1)$$

in which the constant a was assigned different values for the uranium, thorium and actinium series, and the constant b the same value for each. Because of the near-validity of the simple Geiger rule

$$R = cE_{\alpha}^{\frac{3}{2}},$$

connecting the standard range, R, and E_{α}, the initial kinetic energy of a group of a-particles, it was clear that (2.1) could be written, possibly more significantly, in the form

$$\log \lambda = d + f \log E_{\alpha}, \qquad (2.2)$$

and it is this form of the expression which we shall first discuss. Apart from noting that the value of the constant d is not very different for the different series of α-bodies, it will be sufficient to refer to the numerical value of the constant f. If (2.2) is the really significant form of the Geiger-Nuttall relation it implies a power-law connexion between λ and E_{α}. The exponent in this power law is given empirically as $f \sim 80$. Now it is a fact of experience that the significant laws of physics, when expressible as power laws, involve rational exponents of the order of magnitude unity. It is very unlikely, therefore, that (2.2) is the significant form of expression for the undoubted regularity to which Geiger and Nuttall first drew attention. It is an historical fact that other forms of expression were suggested at a relatively early date—thus Swinne (1912, 1913) suggested the form

$$\log \lambda = g + kE_{\alpha}^{\frac{1}{2}}, \tag{2.3}$$

which is fundamentally less objectionable than (2.1) or (2.2)—but, so long as there was no satisfactory theory of α-disintegration, the original form of the empirical regularity was generally adopted by most writers on the subject.

Despite many attempts, nothing approaching a satisfactory theory of α-disintegration was produced until 1928. In that year Gamow (1928) and Gurney and Condon (1928, 1929), independently, put forward the same idea, that of treating the problem of the emission of an α-particle from a radioactive nucleus as a wave-mechanical problem essentially insoluble by the methods of classical mechanics. Previously the opinion had been expressed from time to time—most clearly perhaps by Campbell (1923)—that classical methods might prove inadequate: the development of wave-mechanics, and its immediate success when applied to the evaluation of α-disintegration constants, demonstrated the deep truth that lay behind such tentative expressions of doubt.

If it is true to say that the wave-mechanical treatment of α-disintegration was an immediate success, it is equally true to add that there is, even yet, no real agreement concerning the finer details of the theory. The general features are almost self-evidently 'right'; as to the finer details, it is not so much active disagreement among the experts as the lack of any general method of dealing with the internal dynamics of the atomic nucleus which provides the limitation. As long ago as 1937 Bethe wrote 'Hardly any problem in quantum theory has been treated by so many authors in so many different ways as the radioactive decay', yet quite recently considerable advances in rigour have been achieved in the calculations of the 'penetrability' of a 'potential barrier' (Saha, 1944), and of spin and other effects (Preston, 1947, 1949), which calculations are essential to the theory, without, in fact, touching the more debatable points which refer to the intranuclear dynamics. These points remain almost as obscure as at any time since 1928.

Empirically, the great advance which resulted from the wave-mechanical treatment of α-disintegration was that it provided a natural basis from which the dependence of disintegration probability upon quantities other than available energy could be calculated. There had long been a suspicion that the Geiger-Nuttall rule represented an over-simplification of the essential regularities underlying the phenomenon; Gamow's calculation showed in a qualitative, even in a semi-quantitative, way how the nuclear charge and mass also are determining factors. A clear distinction, however, should be made at this point. The nuclear charge is a determining factor because it fixes the magnitude of the repulsive field over the region in which the force between the α-particle and the residual nucleus is known to be given by the classical coulombian law; the nuclear mass is a factor because it is fairly certain that the effective radius of the nucleus depends chiefly on the mass—and it is this radius which defines the region within which the resultant nuclear field is certainly far from coulombian. Broadly speaking, then, the nuclear charge is a factor the effect of which is fairly closely calculable on the basis of any reasonable approximate theory; the nuclear mass enters in a manner which cannot clearly be foreseen without help from a successful theory of nuclear constitution. On the other hand, general considerations, as well as calculations based on particular hypotheses, would appear to show that over the range of values covered by the α-active bodies of classical radioactivity ($210 \leqslant A \leqslant 238$, $83 \leqslant Z \leqslant 92$) the dependence of λ on Z is likely to be much more important than its dependence on the mass number A. Detailed examination of the experimental results (see § 2.2) will be found to lead to precisely this conclusion.

Before we turn to the experimental results, however, it is important to make one further observation of a general nature concerning the wave-mechanical theory, and to define more closely the simplifying assumptions underlying the particular variant of the theory which we propose to use in discussing the results themselves. The general observation is that the theory in its present form merely provides an approximate means of calculating the disintegration probability when the disintegration energy is known. A complete theory of nuclear dynamics should, of course, make possible the calculation of disintegration energies in terms of universal constants: it need hardly be said that present-day theory is very far from that goal. If the effect of nuclear mass (or radius) upon disintegration probability cannot as yet be determined with any accuracy, it is clear that the more difficult problem of the exact calculation of the binding energies of nuclei is still well beyond the range of attainment. Empirically, certain regularities have long been recognized connecting α-disintegration energies with the mass and charge numbers of the nuclei concerned—in particular the apparent significance of the parameter $3A-4Z$ in this connexion has been noted by several writers (see Wolff, 1921; Fournier, 1929) —but no satisfying qualitative explanation of these regularities is at present forthcoming, much less a detailed numerical theory which should set them in true perspective in the general interpretative scheme.

As a guide in discussing the experimental results we shall adopt, in the first place, the simple 'one body' variant of the wave-mechanical theory as developed by Gamow. In this calculation it is assumed that the α-particle and the residual nucleus can be regarded as distinct entities throughout the whole process of disintegration, and it is further assumed that, when the mass of the residual nucleus has been allowed for, its other properties can be satisfactorily described by a spherically symmetrical force-field which is strictly coulombian 'from infinity' down to the 'inner radius' of the nucleus, r_0. Within this radius it is not necessary to know the form of the force-field with great accuracy, but the assumption is made that at r_0 an abrupt discontinuity occurs in the mutual potential energy of the α-particle and the residual nucleus, the potential energy decreasing, for an infinitesimal decrease in distance, from $2(Z-2)e^2/r_0$ to $-U_0$, and remaining at this latter value for all distances less than r_0. Here Ze is the charge on the nucleus before disintegration. When these assumptions have been made—and a formally plausible, but highly artificial, assumption has been added which relates E_α, U_0 and r_0, and so reduces the number of experimentally undeterminable 'constants' in the final result—we have, according to Gamow, for a disintegration in which the relative motion of the α-particle and residual nucleus is strictly radial throughout,

$$\log\lambda=\log\frac{4h}{mr_0^2}-4\pi(2m)^{\frac{1}{2}}\frac{M-m}{M}\frac{(Z-2)e^2}{h}\frac{1}{E_\alpha^{\frac{1}{2}}}(2u_0-\sin 2u_0),\quad(2.4)$$

where, to sufficient accuracy,

$$\cos^2 u_0=\frac{r_0 E_\alpha}{2(Z-2)e^2}.\quad(2.5)$$

In (2.4) m is the mass of the α-particle and M the mass of the nucleus before disintegration; h is Planck's constant.

Because of the generally small value of $\cos^2 u_0$ we can make the approximation

$$u_0=\frac{\pi}{2}-\epsilon,\quad(2.6)$$

ϵ being a relatively small angle, but, before substituting for ϵ in (2.4) it is useful to note in what way the assumptions of our particular variant of the theory are reflected in the expression as it stands. In doing this we shall find that if we are prepared to accept a 'one body' treatment at all—in order to have a relatively simple result with which to compare the data of experiment—then the detailed assumptions of Gamow's variant of the theory, and the lack of rigour in his mathematics, do not lead to any significant further artificialities in the relation finally obtained for $\log\lambda$. The assumption which removes U_0 is reflected merely in the first term in (2.4); apart from this assumption E_α would occur in this term as well as in the second. But the first term, being itself logarithmic, and the whole variation in E_α of practical importance not being more than a single order of magnitude, whilst λ varies by a factor of 10^{24} or more, it is clear that no important discrepancy results from this assumption. Then, concerning the assumption that the potential energy is strictly coulombian over the range of distances given by $r>r_0$, suffering the discontinuity already specified at $r=r_0$, if it were in fact more correct to say that for $r<r_2$ the potential energy departs significantly from the coulombian (being less than coulombian throughout this range), finally decreasing with r to the value E_α at $r=r_1$, so long as r_0 has been chosen so that $r_1<r_0<r_2$, it will be evident from (2.4) and (2.5) how small a difference this assumption has made. For it will be clear that the difference in question must

be smaller than that resulting from the successive substitution in (2.4) of r_1 and r_2 for r_0—and, when we consider in further detail the approximation indicated by (2.6), we shall appreciate fully the triviality of the latter difference, on any reasonable assumptions about the exact form of the potential energy function appropriate to the ideal 'one body' calculation.

This, then, is the position in respect of Gamow's assumptions—and, concerning the lack of rigour in his mathematical methods, all that need be said, in view of recent criticism, is that the numerical result of that lack of rigour (Saha, 1944) is probably no greater discrepancy than would be represented by a factor of roughly two in λ, approximately constant for all values of E_α. This is not important for our considerations.

Returning now to the approximation represented by (2.6), writing $\cos u_0 = \sin \epsilon = \gamma$, we have

$$2u_0 - \sin 2u_0 = \pi - 4\gamma + \tfrac{2}{3}\gamma^3 \dots .$$

Neglecting third and higher powers of γ we obtain, therefore,

$$\log \lambda = \log \frac{4h}{mr_0^2} - 4\pi^2(2m)^{\frac{1}{2}} \frac{M-m}{M} \frac{(Z-2)e^2}{h} \frac{1}{E_\alpha^{\frac{1}{2}}} + 16\pi m^{\frac{1}{2}} \frac{M-m}{M} \frac{(Z-2)^{\frac{1}{2}} e}{h} r_0^{\frac{1}{2}},$$

or, putting in numerical values, and adopting as units of length (for the measurement of nuclear radius, only) and of energy (for specifying the kinetic energy of the emitted particle) 10^{-12} cm. and 1 MeV., respectively, so that

$$r_0 \text{ cm.} = \rho_0 \times 10^{-12} \text{ cm.}, \quad E_\alpha \text{ erg} = \eta_\alpha \text{ MeV.},$$

$$\log_{10} \lambda = 21{\cdot}60 - 2\log_{10}\rho_0 - 1{\cdot}725 \frac{A-4}{A}(Z-2)\eta_\alpha^{-\frac{1}{2}} + 4{\cdot}08 \frac{A-4}{A}(Z-2)^{\frac{1}{2}}\rho_0^{\frac{1}{2}}. \quad (2.7)\dagger$$

For many purposes we shall see that the total disintegration energy, $E\left(=E_\alpha \dfrac{M}{M-m}\right)$, is more significant than the α-particle kinetic

† If the term in γ^3 had been included, the next term in (2.7) would have been $-2{\cdot}36\left(\frac{A-4}{A}\right)^2 (Z-2)^{-\frac{1}{2}} \rho_0^{\frac{3}{2}} \eta$. For a representative heavy α-emitter the ratio of this term to the last included term in (2.7) is about $0{\cdot}035$.

energy, E_α; if η MeV. is the magnitude of the total energy, clearly the third term in (2.7) becomes (to sufficient accuracy)

$$-1{\cdot}725\frac{A-2}{A}(Z-2)\eta^{-\frac{1}{2}}. \quad (2.8)$$

We shall base subsequent discussion, then, upon (2.7) and (2.8), always remembering the single condition that this expression for $\log_{10}\lambda$ refers only to disintegrations in which the relative motion of α-particle and residual nucleus is strictly radial—that is to disintegrations which do not involve a change in the resultant angular momentum of the nucleus which disintegrates.

2.2. Survey of new data

It has already been pointed out that, as originally drawn, the Geiger-Nuttall diagram focused attention on the individuality of the three disintegration series. Equation (2.7), however, provides no basis for any clear-cut distinction: the effect of nuclear mass-type, which of necessity is constant for any one series,† can be reflected only in small differences in ρ_0, and these can have little effect on the value of $\log_{10}\lambda$. Our object should be, therefore, to exhibit the Geiger-Nuttall regularity without artificial emphasis on the distinctness of the series. It is fortunate that the data of experiment are now much more accurate, and considerably more extensive, than they were when Geiger (1921) last collected together all the information available in a diagram of the conventional type—or even when Gamow and Houtermans (1928) first attempted the analysis of already more detailed experimental results on the basis of the wave-mechanical theory. It is worth while cataloguing the additional sources of information. First, since 1928, the disintegration constant of thorium C′, the shortest-lived α-active body of the three series, has been determined within 2% by direct experiment (Hill, 1948; Bunyan, Lundby and Walker, 1949). This provides an important check on any theory, since it greatly extends the range in which comparison between theory and experiment is possible. Secondly, the range of comparison has been extended in the opposite direction through the improvement of electrical counting

† Since A changes by 4 units in α-disintegration and does not change in β-disintegration, if $4p<A<4(p+1)$ (p, integral), $A-4p$, the integer defining the mass type in this connexion, is unaffected either by α- or by β-disintegration.

methods and the consequent gain in accuracy, both in evaluations of the lifetimes (see Hyde, 1946*a*, 1946*b*; Goldin, Knight, Macklin and Macklin, 1949) of the long-lived bodies, and in determinations of the energies of the α-particles which they emit.† Next, the discovery of α-particle 'fine structure' (Rosenblum, 1929) has added greatly to the sum total of information, and provided clear evidence for the need to extend the interpretative scheme to take count of changes in nuclear spin in certain cases (Gamow, 1930, 1932). Finally, new modes of branching have been discovered in the three series, bringing to light the existence of hitherto unknown 'natural' α-active bodies (see Feather, 1948*a*), and the number of 'artificially produced' α-active bodies is rapidly increasing as the result of the availability of intense sources of neutrons (see Seaborg and Perlman, 1948; Sullivan, 1949; Meinke, Ghiorso and Seaborg, 1951*a*, 1951*b*).

Lacking a considerable amount of this information, Berthelot published a critical survey of the 'energies and periods' in 1942 (Berthelot, 1942*a*, 1942*b*), and independently, in 1944, and in greater detail in 1945, the present writer, being more favourably placed in respect of the new information, discussed the matter—very much from the same standpoint as Berthelot had adopted—in two unpublished reports (Feather, 1944*a*, 1945). Since 1945 other authors have published surveys of available results adopting a very similar standpoint.‡ So far as the Geiger-Nuttall diagram is concerned, the new presentation was precisely the same in the surveys of Berthelot and the writer; $\log_{10} \lambda$ was plotted against η, and the attempt was made to draw smooth curves through the points representative of the disintegrations of species for which Z is constant. The 1945 diagrams have been revised and the most recent experimental results have been incorporated in figs. 2–8, to which reference should be made. In these figures plotted points are labelled with the mass numbers of the various α-active bodies represented on the diagrams, and when 'fine-structure' groups are in question separate points are plotted for the component groups,

† See Clark, Spencer Palmer and Woodward, 1944*a*, 1944*b*, 1945; Cranshaw and Harvey, 1948; Ghiorso, Jaffey, Robinson and Weissbourd, 1949; Wilkinson, 1950.

‡ Vigneron, 1947; Karlik, 1948; Berthelot, 1948; Perlman, Ghiorso and Seaborg, 1948, 1949, 1950; Biswas, 1949*a*, 1949*b*; Jentschke, 1950.

'partial' disintegration constants being calculated on the basis of a knowledge of the relative intensities of the groups. Representative 'points' indicated by ellipses or straight lines refer to disintegrations regarding which there is still some uncertainty concerning disintegration constant and disintegration energy—or at least concerning one of these quantities. The method used in drawing in the smooth curves for the different values of Z may be illustrated from fig. 2 which refers to the α-active isotopes of the elements $Z=84$, 90 and 96. For $Z=84$ conditions are simpler than for any other element; except for one species ($^{211}_{84}$AcC′) no important fine structure has been observed,† and the number of separate species for which accurate information is available is greater for this value of Z than for any other. Disintegration energies increase monotonically as A decreases from 218 to 212 and the points for $A=218$, 216, 215, 214 and 212 fix a smooth curve with considerable accuracy. There is no information concerning the position of the point for $A=217$, and the point for $A=213$ is not far below the curve. Additional comment would be unnecessary, but for the fact that the points for $A=211$ are altogether anomalously placed (Leininger, Segrè and Spiess, 1951)‡, and there are other well-determined points, for $A=210$, 209 and 208, which lie well below any reasonable extrapolation of the smooth curve which has been drawn. These, however, are all points for the lightest isotopes, and for these the previous regular trend of disintegration energy with mass number is decisively reversed, the available energy decreasing very rapidly with A, through $A=211$ to $A=210$, and varying very little thereafter. We use this fact as justification for neglecting these points at this stage, merely remarking that later consideration will show that their positions are, after all, not so anomalous as at first sight they appear.

For $Z=90$, though conditions are not so simple, a general monotonic increase of η with decreasing A again characterizes the

† A γ-radiation of low intensity has been observed with $^{210}_{84}$Po (Bothe and Becker, 1930; de Benedetti and Kerner, 1947; Siegbahn and Slätis, 1947; Alburger and Friedlander, 1951; Grace, Allen, West and Halban, 1951), but so far the corresponding α-group has not been detected (see, however, de Benedetti and Minton, 1952).

‡ It has now been established that AcC′ is an isomeric state of the nucleus ($^{211}_{84}$) (Spiess, 1951). A point for the α-disintegration of the ground-state nucleus is included in fig. 2.

26 representative points for the sequence of mass numbers, $A=232$, 230 (4 points), 229 (3 points), 228 (3 points), 227 (13 points), 226 and 225 and 7 of these points fix a smooth curve which appears to be very closely 'parallel' to the curve which would be chosen as the most obvious extrapolation of the smooth curve for $Z=84$ to lower values of $\log_{10}\lambda$ and η. All the other 19 representative points lie below the curve for $Z=90$, as so fixed. For $Z=90$, we have much evidence of α-particle fine structure, but we again withhold comment, except to remark that it is the natural thing to expect (see §2.3) that when there is a finite change of spin in α-disintegration the disintegration constant will be smaller than would otherwise be the case: we should, for the moment therefore, be undisturbed by points which lie below the smooth curves to which they 'belong', and suspicious only if a representative point is found which lies above its natural curve.

The curves for $Z=84$ and $Z=90$ having been drawn in as described, the remaining smooth curve in fig. 2 (for $Z=96$) was added, without close reference to the six experimental points, so as to constitute, with the curves previously drawn, a family of approximately equally spaced curves of constant form. We may note here the first example of a representative point (for $A=238$), the position of which is uncertain due to lack of knowledge of the contribution of electron capture to the total disintegration rate: other examples of similar uncertainty will be recognized on figs. 5 and 6.

On figs. 3 and 4 the procedure adopted in the construction of fig. 2 has been repeated for $Z=86$ and 92 and for $Z=88$ and 94, respectively, the experimental points for each value of Z being sufficiently numerous to determine a smooth curve with fair accuracy. Reviewing figs. 2–4, which include all the points for elements of even Z, the success of the new presentation is obvious at first sight: no point lies significantly above the curve appropriate to the corresponding value of Z, and in all cases some points—and in some cases most of the points—lie sufficiently closely on the curves to justify their placing with considerable confidence.

Before passing to the consideration of the elements of odd Z it is useful to return to certain matters which have not been fully discussed in the preceding paragraphs but which are important for

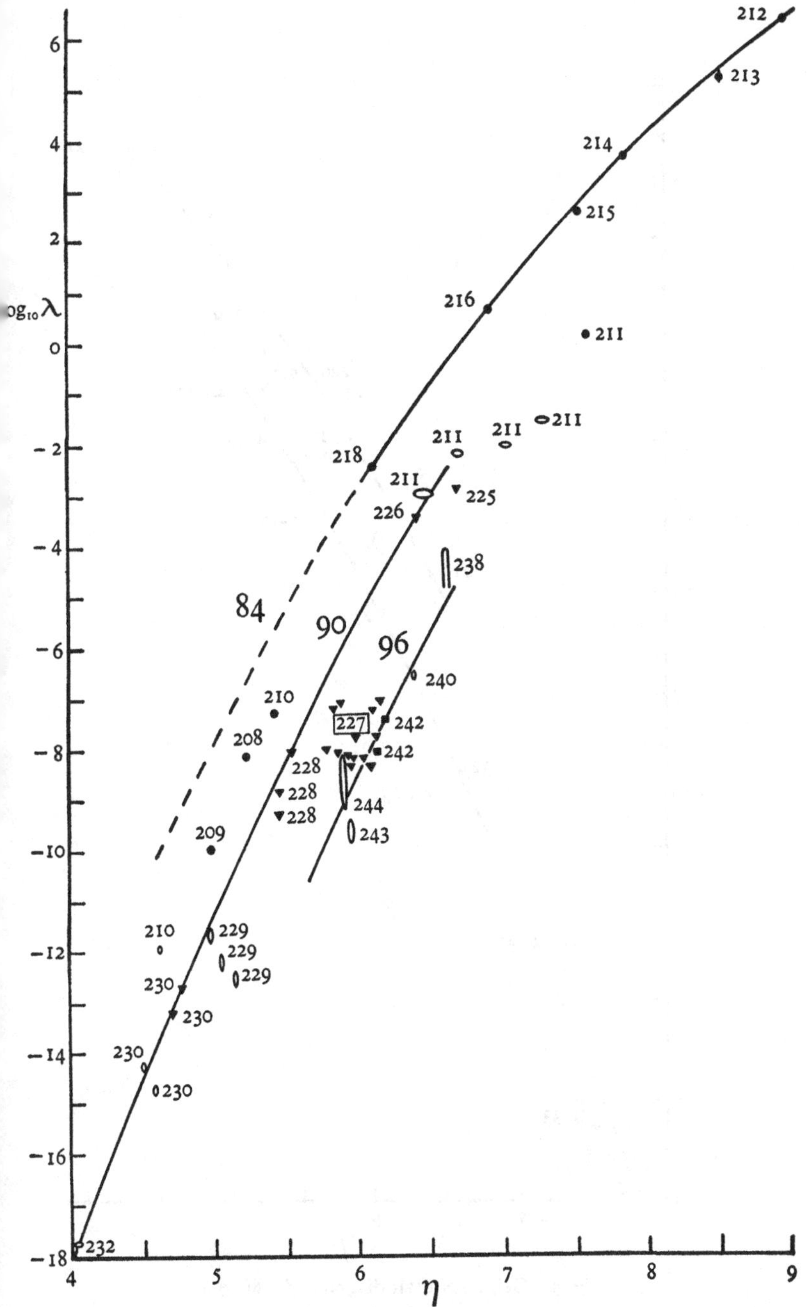

Fig. 2. Geiger-Nuttall diagram, $Z=84, 90, 96$.
λ in sec.$^{-1}$, η in MeV.

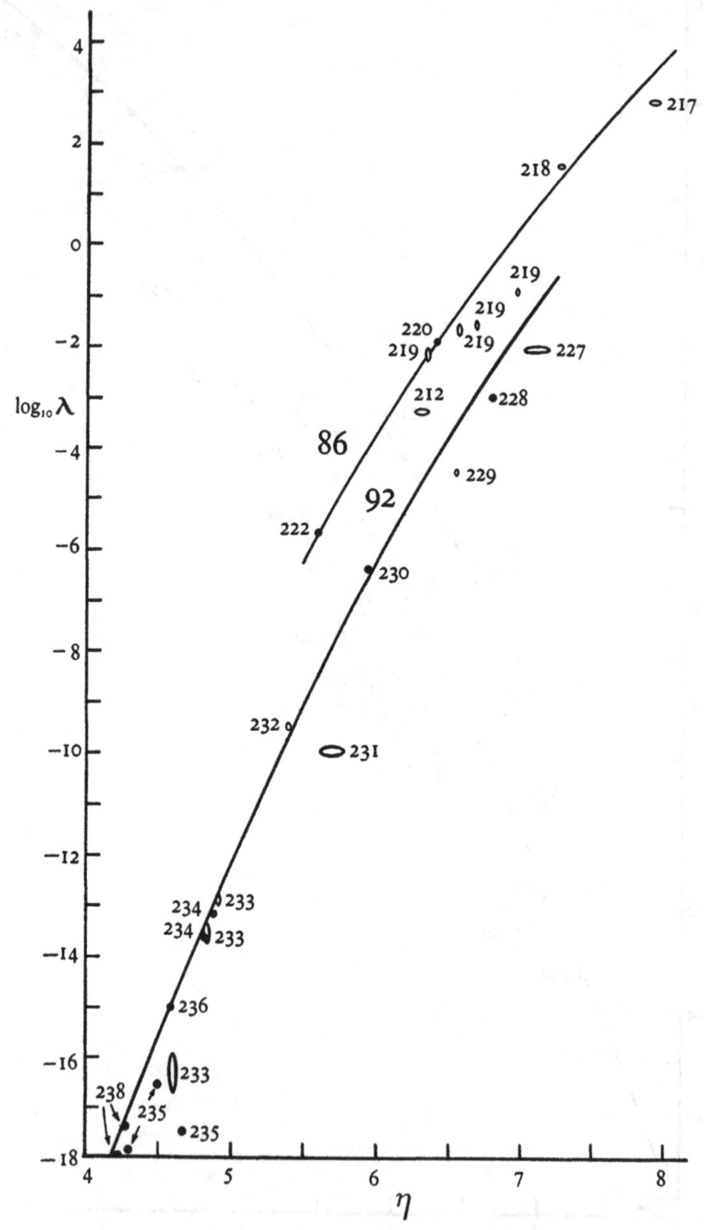

Fig. 3. Geiger-Nuttall diagram, $Z=86$, 92.
λ in sec.$^{-1}$, η in MeV.

our general understanding of the situation. First we should note that our ability to draw smooth curves for Z constant which pass through the representative points as accurately as do the curves of

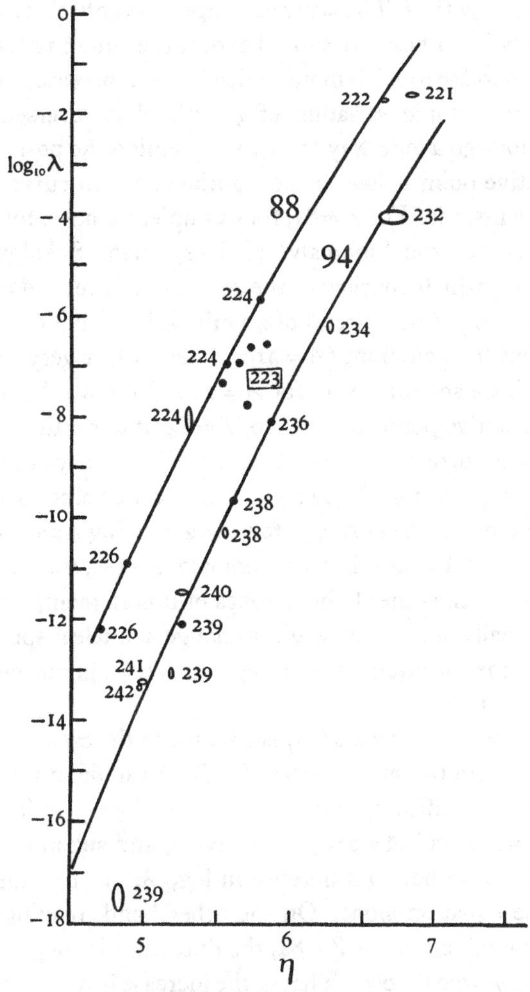

Fig. 4. Geiger-Nuttall diagram, $Z=88$, 94. λ in sec.$^{-1}$, η in MeV.

figs. 2–4 (for those points which we assume to refer to disintegrations without change of spin)† can only be understood in the light of

† Strictly, disintegrations for which both initial and final nuclear states have zero angular momentum—and, to a good approximation, for any disintegration in which these states have the same angular momentum (see p. 53).

equation (2.7) when the monotonic variation of η with A is regarded as significant, and when at the same time it is assumed that the nuclear characteristic represented by ρ_0 in the equation also varies monotonically with A. The latter assumption is entirely reasonable: on any likely hypothesis we should expect the 'inner radius' of the nucleus to increase roughly monotonically, as A increases. Secondly, when the monotonic variation of η with A is reversed, similar considerations go a long way towards explaining the position of the representative point below the appropriate smooth curve as extrapolated downwards. For $Z=84$, for example, we have, for the time being, to accept the 'anomaly' (Fajans, 1926; Sokolow, 1927; Meyer, 1932) which consists in the reversal, between $A=212$ and $A=211$, of the previous trend of η with A, but having accepted it we note that for polonium ($A=210$) η has in fact very nearly the value which we should expect for $A=220$. Now we should expect the representative point for $A=220$, $Z=84$ (if our estimate of η for this species is correct) to lie on the extrapolated curve of fig. 2. In that case the point for $A=210$ (with the same value of η) would definitely lie below the curve, ρ_0 for $A=210$ being almost certainly smaller than for $A=220$. In this 'anomalous' case, then, and in the similar cases of the other light isotopes of this element, it may well be that we shall not need to assume a change of nuclear spin in order to explain the position of the representative point below the appropriate curve.

The argument may be made quantitative in the case of polonium as follows. From the smooth curve for $Z=84$ in fig. 2 we can take, as nearly as is significant, $\eta=9{\cdot}0$ ($\eta^{\frac{1}{2}}=3{\cdot}0$) and $\eta=5{\cdot}76$ ($\eta^{\frac{1}{2}}=2{\cdot}4$) to refer to $A=212$ and $A=219$, respectively, and substituting these values in (2.8) we have a difference in $\log_{10}\lambda$, of $11{\cdot}73$ due to this term in the equation alone. On the other hand, reading directly from the smooth curve for $Z=84$, the difference in $\log_{10}\lambda$ between $\eta=5{\cdot}76$ and $\eta=9{\cdot}0$ is $10{\cdot}35$. That is, the increase in λ with increasing η is less rapid than is given by (2.8) only—and for a difference in mass number of 7 units $(219-212)$ the deficit in $\log_{10}\lambda$ is $1{\cdot}38$. Now the point for polonium is $1{\cdot}56$ in $\log_{10}\lambda$ below the smooth curve—and we have already suggested that this deficit is to be correlated with a difference in mass number of 10 units $(220-210)$. Assuming a linear correlation, the deficit for $\Delta A=7$ would then be $1{\cdot}09$ (to be compared with 1·38 above deduced). The agreement is sufficiently good, and two points may be noted concerning it: the first is that, if the agreement is significant at all, then it must be assumed that (2.8) represents with considerable accuracy the actual dependence of $\log_{10}\lambda$ on η for Z constant—and the second is that it is not necessary to interpret the nuclear characteristic responsible for the 'η-independent' effect in terms of a nuclear radius ρ_0 as implied in (2.7), so long as the characteristic which is responsible for the effect is linearly dependent upon A, for small changes in A.

If we wish to interpret the 'η-independent' effect in terms of the inner radius ρ_0 we can submit our interpretation to numerical test as follows. Using (2.8) we re-write equation (2.7) in the form

$$\log_{10}\lambda = L - 1{\cdot}725\frac{A-2}{A}(Z-2)\eta^{-\frac{1}{2}}. \quad (2.9)$$

Then we have, for Z constant, and assuming $\rho_0 = RA^{\frac{1}{3}}$,

$$\frac{\partial L}{\partial A} = -\frac{1}{3{\cdot}453A} + 0{\cdot}68R^{\frac{1}{2}}A^{-\frac{5}{6}}\frac{A+20}{A}(Z-2)^{\frac{1}{2}}. \quad (2.10)$$

Accepting as an experimental result for $Z=84$, $\partial L/\partial A = 0{\cdot}156$† at a mean value of A of 215 (see above), (2.10) allows us to calculate a value for R. The plausibility of our interpretation then depends upon the numerical value obtained—and particularly upon the relation between the value so obtained for R and the value which may be calculated by solving (2.9), with $\rho_0 = RA^{\frac{1}{3}}$, in respect of a representative case. As such a case we may take $\log_{10}\lambda = 2{\cdot}00$, $\eta = 7{\cdot}29$ ($\eta^{\frac{1}{2}} = 2{\cdot}7$), $A = 215$, $Z = 84$ (see fig. 2).‡ The results of the two calculations so made are:

from (2.9) $R = 0{\cdot}131$
from (2.10) $R = 4{\cdot}2$ (unit of $R = 10^{-12}$ cm.).

At first sight it would appear that the large discrepancy which our calculations reveal would be decisive against the interpretation under test, but further consideration modifies an over-hasty judgement. The calculation from (2.9) indicates that the effective

† This neglects the variation of the second term in (2.9) with A. To take this term into account would involve adding roughly 0·002 to the value here accepted for $\partial L/\partial A$.

‡ Actually the true disintegration energy in the case of ${}^{215}_{84}$AcA is 7·51 MeV. ($\log_{10}\lambda = 2{\cdot}58$); however, we take the idealized values as more simply representative of the $Z=84$ curve as a whole.

inner radius of the mid-sequence nucleus $\binom{215}{84}$ is $7{\cdot}83 \times 10^{-13}$ cm.;† that based upon (2.10) is discordant with this result only in so far as it implies that the increase of effective radius with A amongst members of the isotopic series $\binom{210}{84}$ to $\binom{218}{84}$ is considerably more rapid than corresponds with the general increase of ρ_0 with A amongst the stable nuclei. A radius of $7{\cdot}83 \times 10^{-13}$ cm. for a nucleus of mass number 215 is entirely in line with what would be expected on the basis of other experimental results; if we accept this value, further calculation involving $\partial L/\partial A$ merely indicates that the increase of ρ_0 consequent upon unit increase in A (Z constant) about $A=215$ is some 1·01 %, whereas it would be no more than 0·155 % if the result $\rho_0 = RA^{\frac{1}{3}}$ were valid for the heaviest nuclei (as it would appear to be fairly closely valid in general for all nuclei but the heaviest).

So far we have dealt only with the α-active species of even Z; for those of odd Z the experimental results are so much less extensive that a somewhat different treatment is necessary. Figs. 5–7 contain the experimental points for species having $Z=85, 87, 89, 91, 93$ and 95, plotted on the same scale and indicating probable errors of determination in precisely the same way as the 'points' on figs. 2–4. It will be clear from inspection that for no odd value of Z are the points sufficiently numerous to determine unambiguously the appropriate curve, thus smooth curves have been drawn in, as some of the curves of figs. 2–4 were drawn, by interpolation. The method of interpolation was to draw the curves for odd Z in such a way that in a composite diagram (fig. 9) covering all values of Z from 84 to 96 (both values inclusive),‡ these curves would lie mid-way between

† If (2.7) and (2.9) were carried to one further stage of approximation (see footnote, p. 32) calculated values of ρ_0 would be increased by approximately $\frac{125}{Z-2}\eta\rho_0$ %. The increase in the present case would thus be 8·7 %. The general conclusions of the present section are, however, qualitatively independent of the fact that our calculated values of ρ_0 do not include this small (variable) correction.

‡ Experimental results concerning the α-active isotopes of berkelium ($Z=97$) and californium ($Z=98$) have not been included in our survey. Before long, however, such results should be sufficiently accurate and numerous for inclusion (see Thompson, Ghiorso and Seaborg, 1950; Thompson, Street, Ghiorso and Seaborg, 1950; Ghiorso, Thompson, Street and Seaborg, 1951; Hulet, Thompson, Ghiorso and Seaborg, 1951).

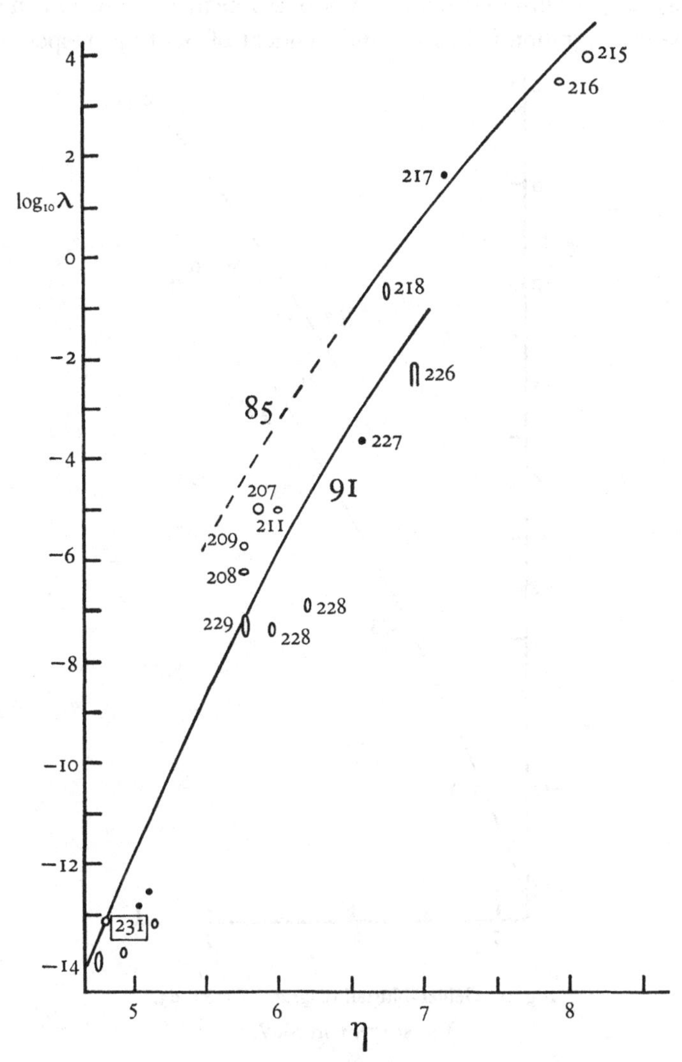

Fig. 5. Geiger-Nuttall diagram, $Z=85, 91$.
λ in sec.$^{-1}$, η in MeV.

the curves for the adjacent even-numbered elements already given in figs. 2–4. To proceed in this way is to assume that the phenomenon of α-disintegration is broadly independent of parity in respect of

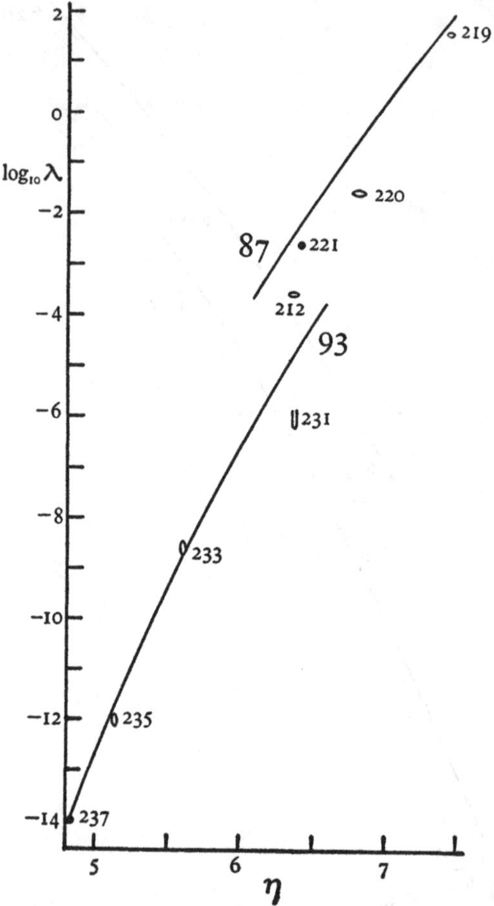

Fig. 6. Geiger-Nuttall diagram, $Z=87, 93$.
λ in sec.$^{-1}$, η in MeV.

Z—and the general success of the procedure is obvious at once from the separate figures. Since we should expect that spin changes in α-disintegration would be more frequent amongst the active species of odd Z than is the case for even Z, we should expect relatively more points to lie below the appropriate curves in figs. 5–7 than in figs. 2–4, but it is significant that in the former figures there are points lying on the curves, also—and it is important to note that there are in fact no points lying appreciably above the curves to which they should belong. Both observations emphasize the general success which has been claimed.

Figs. 2–7, then, represent the complete Geiger-Nuttall diagram, except that the representative points for the α-active species having $Z=83$ are not included. These are plotted separately in fig. 8. Taken by themselves the points on fig. 8 show no abnormality: they bear witness to well-marked fine structure in the α-particle spectra

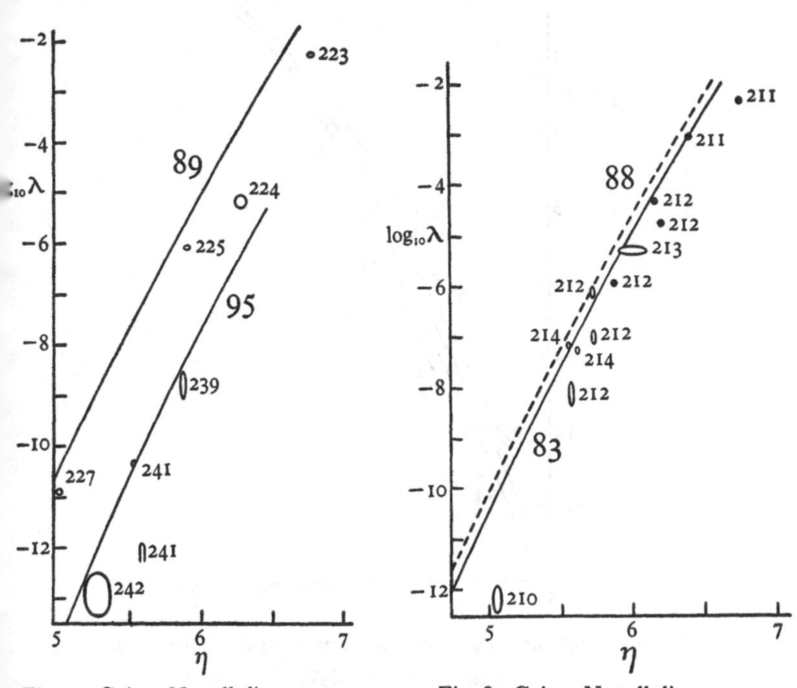

Fig. 7. Geiger-Nuttall diagram, $Z=89, 95$. λ in sec.$^{-1}$, η in MeV.

Fig. 8. Geiger-Nuttall diagram, $Z=83$. λ in sec.$^{-1}$, η in MeV.

of $^{214}_{83}$Ra C, $^{212}_{83}$Th C and $^{211}_{83}$Ac C, but a curve of the standard form can be drawn so as to pass through a number of points, leaving none above the curve. Also the maximum α-disintegration energy increases step by step as the mass number decreases from 214 to 211. Radium E (mass number 210) in its recently discovered rare mode of α-disintegration (Broda and Feather, 1947) shows the 'polonium anomaly' very clearly and requires no further consideration on that account.† But the whole curve for $Z=83$ is anomalous in position when considered in relation to the curves of the earlier figures.‡ For sake of comparison the curve for $Z=88$, taken from fig. 4, is shown dotted in fig. 8. It is obvious at first sight that the regularity which has been so marked over the whole range $96 \geqslant Z \geqslant 84$ (see fig. 9) is decisively broken for $Z=83$, and that disintegration con-

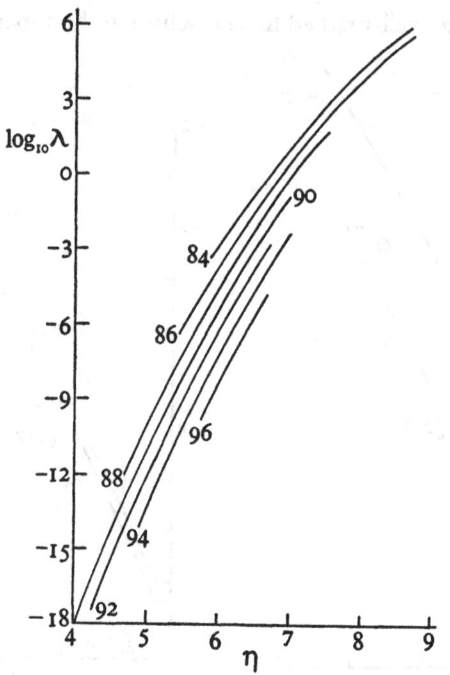

Fig. 9. Composite Geiger-Nuttall diagram, $84 \leqslant Z$ (even) $\leqslant 96$. λ in sec.$^{-1}$, η in MeV.

stants for α-active species having this value of Z are consistently smaller (by a factor of about 180) than might otherwise have been expected. If we attempt to interpret this result in terms of the 'one-body' theory, by calculating an effective inner radius as has

† A second α-active state of RaE has recently been identified (Neumann, Howland and Perlman, 1950; Levy and Perlman, 1951), but its lifetime is not sufficiently well known for a point to be plotted on fig. 8 (see Feather, 1951).

‡ Consideration of the representative points for the recently discovered neutron-deficient α-active bismuth isotopes is deferred until later (p. 51).

been done previously, we have, for the near-representative isotope $Z=83$, $A=210$,†

$$\eta=4{\cdot}87\pm0{\cdot}05,\quad \log_{10}\lambda=-12{\cdot}2\pm0{\cdot}3,\quad \rho_0=0{\cdot}626\pm0{\cdot}019.$$

This value has to be compared with the value $\rho_0=0{\cdot}737$ for the neighbouring isobar $\binom{210}{84}$. The discrepancy is quite startling‡: on the basis of this crude interpretation it must be supposed that the effective inner radius for α-disintegration increases abruptly by almost 18 % between these isobaric active species having $Z=83$ and $Z=84$, whilst for each increase of one unit in Z thereafter (up to $Z=96$) the increase (between 'comparable' species) is, on the average, 1·0 % (see fig. 10(*a*) for details). Empirically, there is a very clear abnormality which is self-evidently a charge-dependent effect; the only doubt concerning its interpretation in terms of an abrupt change in radius is the magnitude of the change which, on the simple one-body theory, is required. We might explore the possibility of an alternative explanation postulating the pre-existence of α-particles in all nuclei of charge number greater than 83—and the need for their formation out of neutrons and protons in nuclei of smaller charge—but until we have examined the experimental evidence from another angle such discussions would appear to be premature.

Before proceeding to this examination, however, our analysis in respect of ρ_0 may be carried one stage farther. In order to do this we have to introduce a new concept, that of the 'representative isotope'. This concept will be considered more systematically in § 2.6 below, here we need only state that for each value of Z from $Z=84$ to $Z=96$ there exists at least one active isotope which is stable both against β- and against capture-disintegration. These species are 'pure α-emitters', and they represent the continuation of the system of 'stable' species beyond $Z=83$. There are never more than two pure α-emitters of odd A for any value of Z. If there is only one such species in any case, then we take this as the representative isotope of that element; if there are two pure α-emitters of odd A for a given Z, then the even-A isotope intermediate in mass

† See p. 48 below.

‡ A more recent estimate of the α-disintegration energy of $^{210}_{83}$RaE gives $\eta=5{\cdot}06\pm0{\cdot}03$, $\rho_0=0{\cdot}573\pm0{\cdot}015$—increasing the discrepancy here noted.

between them is the representative isotope. We indicate the mass number of the representative isotope, so identified, by $A_r(Z)$. Clearly, when ρ_0 varies rapidly with A, for Z constant, it is useful to have an objectively chosen value of A, such as $A_r(Z)$, in terms of which to follow the essential variation of ρ_0 with Z. Table III gives the mass numbers of the representative isotopes for $83 \leqslant Z \leqslant 96$, and fig. 10($a$, b) the values of ρ_0 for these isotopes, exhibited both as a function of Z (fig. 10a) and also as a function of $A_r(Z)$ (fig. 10b).

TABLE III

Z	83	84	85	86	87	88	89	90	91	92	93	94	95	96
$A_r(Z)$	209	212	215	217	219	222	225	228	231	234	237	239	241	(244)

In calculating the values of ρ_0 for fig. 10 the following rules have been observed: (i) no point has been plotted on fig. 10 for any Z (except for $Z=83$, 86, 94) unless at least one point representing one of the modes of α-disintegration of the representative isotope for that value of Z lies effectively on the appropriate smooth curve of the revised Geiger-Nuttall diagram (figs. 2–8); (ii) when more than one point for $A_r(Z)$ satisfies this acceptance test, that point corresponding to the α-mode of greatest energy release has been used as giving values of η and λ for the derivation of ρ_0 from (2·9) and (2·7) above; (iii) for $Z=83$, ρ_0 calculated for the α-disintegration of $^{210}_{83}$Ra E has been decreased by 1 % to give ρ_0 for the representative (stable) isotope $^{209}_{83}$Bi, for $Z=86$ a similar decrease has been applied to ρ_0 calculated for the α-disintegration of $^{218}_{86}$Em to give ρ_0 for $^{217}_{86}$Em, and for $Z=94$ calculations based on the α-disintegration of $^{238}_{94}$Pu have been used to give ρ_0 for $^{239}_{94}$Pu by appropriate adjustment. The result of the application of these rules is that no point has been plotted on fig. 10 either for $Z=85$ or for $Z=89$, that the point for $Z=91$ has been derived from calculations relative to the α-disintegration mode of $^{231}_{91}$Pa which leads to the fourth excited state of the daughter nucleus $^{227}_{89}$Ac, that the points for $Z=92$ and $Z=95$ similarly refer to what are probably α-disintegrations to first excited states of the respective daughter nuclei, and that the remainder of the points, with more or less certainty in individual cases, refer to ground-to-ground state disintegrations. Owing to the restrictions imposed by the rules of acceptance, it may reasonably be assumed that all the plotted points refer to disintegrations in which the change in nuclear angular momentum is zero.

The full curves in fig. 10(a, b) have been drawn with reference only to the points for even values of Z. The points for odd Z in general lie below these curves: there is the wholly abnormal drop, representing some 17 % of ρ_0, between the points for ${}^{212}_{84}$Th C′ and

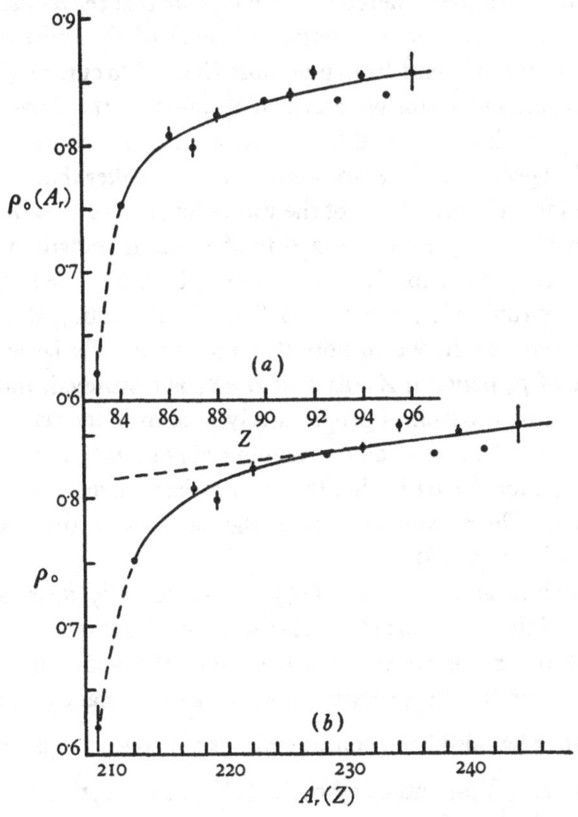

Fig. 10. Effective nuclear radius for representative isotope, (a) as a function of Z, (b) as a function of $A_r(Z)$. ρ_0 in 10^{-12} cm.

${}^{209}_{83}$Bi, and the points for ${}^{219}_{87}$Fr, ${}^{237}_{93}$Np and ${}^{241}_{95}$Am are somewhat more than 1 % low (in fig. 10b). Only the point for ${}^{231}_{91}$Pa appears as an exception to this trend, but this exception may well be more apparent than real. It has been stated that an α-group of low energy (0·33 MeV. less than normal) has been used in fixing the position of this point, and reference to fig. 5 will show that the effect of this is that we start from a point on the Geiger-Nuttall curve which would be more appropriate to the ground-to-ground state disintegration of $^{232}_{91}$Pa than to that of $^{231}_{91}$Pa. In this case we should expect to obtain a ρ_0 value which would be some 1 % too high for the lighter isotope. This is precisely the order of the discrepancy in question. To sum up, therefore, we conclude that the α-disintegrations (of zero angular momentum change) of the representative isotopes of the elements from polonium ($Z=84$) to curium ($Z=96$), inclusive, are satisfactorily described by values of the inner nuclear radius ρ_0 which vary smoothly with $A_r(Z)$, except that the values of ρ_0 for the species of odd Z are some 1–1·5 % smaller than would be expected from interpolation of the values for species of even Z. The trend of the latter from $A_r=230$ to $A_r=244$ is indistinguishable (see fig. 10*b*) from the 'regular' trend given by $\rho_0=RA^{\frac{1}{3}}$, with $R=0{\cdot}137$† (unit of ρ_0 and $R=10^{-12}$ cm.). Accepting this as the regular trend we may also note that the abnormally large rate of increase of ρ_0 between $Z=83$ and $Z=84$ is continued, though to a degree which becomes progressively less, over the whole range $84 \leqslant Z \leqslant 90$. If there is any remaining abnormality it is that the value of ρ_0 for $Z=92$ is slightly greater than might otherwise be predicted. There may be some significance in the fact that $92=84+8$ (see p. 26).

Rosenblum and Valadares (1950) have recently discussed the problem of the ρ_0 values of the heavy α-active elements disregarding some of the refinements introduced into the above discussion. Broadly their conclusions are not seriously at variance with ours. The matter has also been considered, for $\binom{e}{e}$ species separately, by Perlman and Ypsilantis (1950) and Kaplan (1951). An earlier discussion is that of Preston (1946).

It is at this stage in the argument from fig. 10 that some very rough results, not hitherto taken into consideration, become of cardinal

† See footnote, p. 42, regarding the 'absolute' value of this constant. A refinement introduced by Ambrosino and Piatier (1951) indicates that the 'true' value of ρ_0 is slightly less than that calculated on the one-body theory of Gamow, to whatever degree of approximation (2.7) is carried. In Gamow's theory the lowering of the potential barrier due to the field of the extranuclear electrons is neglected, and the fact that some disintegration energy is 'lost' in the adiabatic reorganization of these electrons (p. 9) is 'forgotten'. Ambrosino and Piatier claim that these two effects are independent and that to take count of them in (2.7) would involve increasing η by about 70 keV. for a representative heavy α-emitter (see Rasmussen, Thompson and Ghiorso, 1951).

importance. The first of these results concerns an α-active isotope of gold of mass number less than 190, and probably between 185 and 187, which has been found by Thompson, Ghiorso, Rasmussen and Seaborg (1949). This species transforms principally by K capture, and only to the extent of about 0·01 % by α-disintegration. The half-value period is 4·3 min. and the α-disintegration energy 5·3 MeV. An effective value of ρ_0 of $0{\cdot}667 \pm 0{\cdot}023$ is obtained from (2.9) and (2.7). This is already greater than the value ($0{\cdot}620 \pm 0{\cdot}019$) plotted in fig. 10 for $^{209}_{83}$Bi, and it is as likely as not to be somewhat less than the 'correct' value for the species concerned just because of the frequent incidence of finite spin changes in α-disintegrations of species of odd Z. Also, the α-active species concerned is very considerably lighter than the representative (stable) isotope of gold $^{197}_{79}$Au. Assuming the type of variation of ρ_0 with A which we have found general with the heavier elements, we should conclude that, for $^{197}_{79}$Au, $\rho_0 \geqslant 0{\cdot}72 \pm 0{\cdot}04$. This result shows up the low value of ρ_0 for the representative isotope of $Z=83$ in even clearer perspective than before: ρ_0 for the representative isotope obviously increases again as Z decreases (for a small range at least) below this value. The increase may well be to the 'normal' value given by

$$\rho_0 = 0{\cdot}137\{A_r(Z)\}^{\frac{1}{3}};$$

information at present available is quite inadequate to determine the issue.

A second set of rough results refers to certain low-mass isotopes of the elements from bismuth (83) to francium (87).† In many cases these recently discovered species transform predominantly by electron capture, and the branching ratios for α-disintegration are only imperfectly known. In many cases, too, the question of α-particle fine structure is still an open question. But where the experimental evidence appears reasonably clear-cut it already provides important information. We shall regard the evidence as reasonably clear-cut for the species $^{212}_{87}$Fr, $^{212}_{86}$Em, $^{211}_{85}$At, $^{208}_{85}$At, $^{209}_{84}$Po, $^{208}_{84}$Po and (though admitting wider limits of uncertainty) $^{203}_{85}$At, $^{207}_{84}$Po, $^{206}_{84}$Po, $^{205}_{84}$Po, $^{201}_{83}$Bi, $^{199}_{83}$Bi, $^{198}_{83}$Bi. We tabulate below the effective values of ρ_0 calculated for the α-disintegrations of these species.

† Howland, Templeton and Perlman, 1947; Templeton and Perlman, 1948; Ghiorso, Meinke and Seaborg, 1949; Hyde, Ghiorso and Seaborg, 1950; Neumann and Perlman, 1950; Karraker and Templeton, 1950; Barton, Ghiorso and Perlman, 1951.

With few exceptions the values of ρ_0 in Table IV are 'effective' values just because it is not certain, in most cases, that the disintegration is characterized by zero spin change; as already mentioned, if this is not so the true value of ρ_0 will be greater than the tabulated value. On the other hand, the true value may be less than the tabulated only if the α/capture branching ratio has been over-estimated in any case. Values of the neutron number N are included in the table.

TABLE IV

Z	87	86	85	85	85	84	84
A	212	212	211	208	203	209	208
ρ_0	0·758 ±0·006	0·765 ±0·008	0·728 ±0·004	0·720 ±0·005	0·781 ±0·005	0·734 ±0·003	0·746 ±0·003
N	125	126	126	123	118	125	124

Z	84	84	84	83	83	83
A	207	206	205	201	199	198
ρ_0	0·733 ±0·019	0·778 ±0·019	0·797 ±0·031	0·712 ±0·014	0·672 ±0·013	0·634 ±0·012
N	123	122	121	118	116	115

In fig. 11 values of ρ_0 are plotted for all isotopes of polonium (84) and bismuth (83) for which available information is sufficiently precise, as also for the astatine isotopes $^{217}_{85}At$ and $^{203}_{85}At$. Supposedly 'true' values of ρ_0 are differentiated from 'effective' values by the type of point used in plotting, and limits of error are shown. Taking the points for $Z=84$ and $Z=83$ together, it is clear that the 'parallel' smooth curves drawn for these two values of Z represent an entirely plausible systematization of the results, and the two points for $Z=85$ might well be accounted for by a curve of similar form for this element, also. If further experimental evidence goes to confirm this speculative construction then it will be clear that the monotonic variation of ρ_0 with A (Z constant) noted in the analysis of earlier results does not hold when heavy α-active species having $N \leqslant 126$ are in question. For nuclei having this value of the neutron number, α-disintegration appears to be intrinsically a more 'difficult' process than it is for nuclei containing any number of neutrons greater than 114, or thereabouts. In the range $114 \leqslant N \leqslant 126$ a minimum in the intrinsic difficulty of α-emission appears to occur at about $N=120$.

There is one final remark about the discussions of this section which must now be made. We have throughout used mass, charge and neutron numbers of the parent species in quoting the values of inner nuclear radius ρ_0 calculated on the one-body theory of α-disintegration. Strictly this theory attributes the effective radius to the daughter nucleus in each disintegration. Equally certainly, however, the detailed theoretical picture must be somewhat vague on just this point. Therefore, whilst we are convinced of the significance of the regularities (and abnormalities) in ρ_0 which our

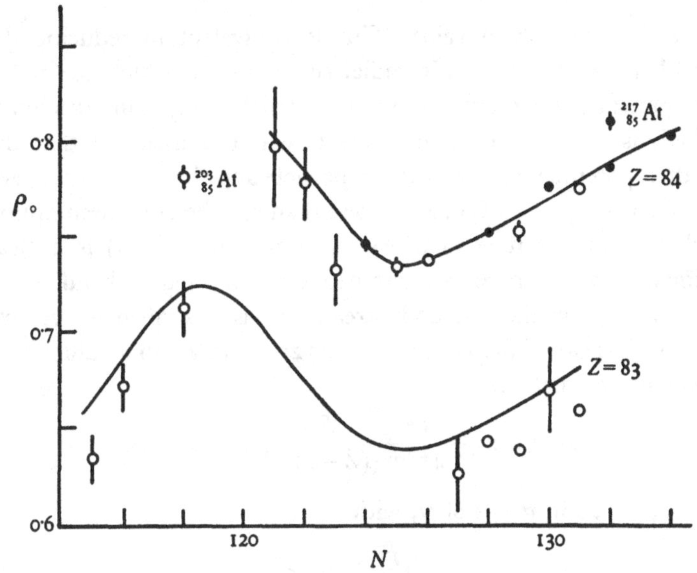

Fig. 11. Effective nuclear radius for isotopes of polonium ($Z=84$) and bismuth ($Z=83$).

ρ_0 in 10^{-13} cm.

● true value, ○ effective value.

approximate treatment has revealed, we leave for later consideration (see p. 62) the problem of referring the abnormalities to one or other nucleus (parent or daughter) involved in the anomalous disintegrations.

2.3. The effect of spin changes

The treatment of the effect of a change in nuclear spin in α-disintegration in the one-body theory of Gamow is even more artificial than is the rest of the theory (Preston, 1947; Devons, 1949; Dancoff, 1950). If angular momentum to the extent of j quantum units (or more†) is taken up in the relative motion of the α-particle and the residual nucleus it is assumed that the disintegration probability can be calculated as before, provided that the mutual potential energy of the two particles, at any distance $r(>r_0)$, is imagined as increased by an amount

$$\frac{h^2}{8\pi^2 m r^2} j(j+1)$$

over the coulombian value. This is equivalent to reducing the problem to one of strictly radial separation in which the radial component of the momentum of the particles only is in question—which is plausible enough—except that it is assumed that the mutual potential energy of the α-particle and the residual nucleus remains $-U_0$ for all values of r less than r_0. The latter assumption (whether or not it is to prove practically important) is a clear admission of failure properly to define the problem in hand.

If we follow the one-body treatment as described above, we obtain, instead of the factor $2u_0 - \sin 2u_0$ in (2.4), to a sufficiently good approximation,

$$2u_0 - \sin 2u_0 + \frac{h^2}{4\pi^2 m} \frac{E_\alpha}{4(Z-2)^2 e^4} j(j+1) \tan u_0,$$

or, expressed in terms of γ, with

$$\gamma^2 = \frac{r_0 E_\alpha}{2(Z-2)e^2} \quad \text{(cf. p. 32)},$$

$$\pi - 4\gamma + \frac{j(j+1)h^2}{8\pi^2 m r_0 (Z-2)e^2}\gamma + \tfrac{2}{3}\gamma^3 \ldots,$$

instead of $\pi - 4\gamma + \frac{2}{3}\gamma^3 \ldots$.‡ The final result is to add to (2.7) the terms

$$-0{\cdot}372(Z-2)^{-\frac{1}{2}} \rho_0^{-\frac{1}{2}} j(j+1) - 2{\cdot}36 \left(\frac{A-4}{A}\right)^2 (Z-2)^{-\frac{1}{2}} \rho_0^{\frac{3}{2}} \eta. \quad (2.11)$$

† If the angular momenta of initial and final nuclear states are I_1 and I_2 quantum units, respectively, then the angular momentum taken up in relative motion may have any one of the values $|I_1 - I_2|$, $|I_1 - I_2| + 1$, ..., or $I_1 + I_2$, in the same units.

‡ In this section we include the term in γ^3, previously omitted (p. 32), since for the heavy α-active bodies, the new term in γ is on the average some $20/j(j+1)$ times smaller than the following term in γ^3, but the inclusion does not in fact make much difference to numerical results, as is made clear in the footnote, p. 42.

Whilst remaining extremely sceptical of the significance of the numerical factor in the new term in j, we shall attempt an interpretation of the experimental data by adopting the form of the result given by the one-body treatment, assuming that the empirical expression for $\log_{10}\lambda$ may be written (cf. (2.9))

$$\log_{10}\lambda = L - 1{\cdot}725\frac{A-2}{A}(Z-2)\eta^{-\frac{1}{2}} - C(Z-2)^{-\frac{1}{2}}\rho_0^{-\frac{1}{2}}j(j+1) - 2{\cdot}36\left(\frac{A-4}{A}\right)^2(Z-2)^{-\frac{1}{2}}\rho_0^{\frac{3}{2}}\eta. \quad (2.12)$$

Here L is a function of ρ_0, A and Z (p. 32) which we assume to be constant for all the disintegration modes which contribute to the fine-structure spectrum of a given α-emitter. If, then, $\lambda_1, \lambda_2, \ldots$, are the partial α-disintegration constants for such a body, if $\eta_1, \eta_2, \ldots$, are the corresponding disintegration energies, and $j_1, j_2, \ldots$, the values of the dominant angular momentum changes concerned (in quantum units) we have

$$\log_{10}\left(\frac{\lambda_1}{\lambda_r}\right) + 1{\cdot}725\frac{A-2}{A}(Z-2)(\eta_1^{-\frac{1}{2}} - \eta_r^{-\frac{1}{2}}) + 2{\cdot}36\left(\frac{A-4}{A}\right)^2(Z-2)^{-\frac{1}{2}}\rho_0^{\frac{3}{2}}(\eta_1 - \eta_r) = C(Z-2)^{-\frac{1}{2}}\rho_0^{-\frac{1}{2}}(j_r - j_1)(j_r + j_1 + 1). \quad (2.13)$$

Writing, ${}_1\Delta_r(Z-2)^{-\frac{1}{2}}\rho_0^{-\frac{1}{2}}$ for the left-hand member of the above equation we obtain

$${}_1\Delta_r = C(j_r - j_1)(j_r + j_1 + 1). \quad (2.14)$$

We adopt the convention that $\eta_1 > \eta_2 > \ldots$, then Tables VI and VII give the experimental values of ${}_1\Delta_r$ for the various (partial) modes of α-disintegration represented in figs. 2–8. The values of ρ_0 given in these tables and used in the calculation of ${}_1\Delta_r$ are greater than those plotted in fig. 10, since the correction for the previously omitted term is now included (footnote, p. 42), but the general results are not very sensitive to ρ_0. For purposes of interpretation, Table V gives the (even integral) values expected for ${}_1\Delta_r/C$ on the basis of (2.14), for different possible combinations of values of j_1 and j_r.

The results for ${}_1\Delta_r$ are set out in two tables according to the following scheme of arrangement: in Table VI are included the results relative to the fine-structure groups of those α-emitters for which the disintegration mode of greatest energy in each case is represented by a point lying on the appropriate Geiger-Nuttall curve in one of figs. 2–8; Table VII includes the results relative to the other α-emitters exhibiting fine structure. By this subdivision Table VI refers to cases in which the ground-to-ground state

TABLE V

j_1 \ j_r	0	1	2	3	4	5	6
0	0	2	6	12	20	30	42
1	−2	0	4	10	18	28	40
2	−6	−4	0	6	14	24	36
3	−12	−10	−6	0	8	18	30
4	−20	−18	−14	−8	0	10	22
5	−30	−28	−24	−18	−10	0	12
6	−42	−40	−36	−30	−22	−12	0

TABLE VI

Z	A	ρ_0	η_r	${}_1\Delta_r$	j_r
96	242	0·896	6·182	0	0
			6·132	2·8 ± 0·4	(2)
94	238	0·898	5·592	0	0
			5·550	3·7 ± 0·7	2
92	238	0·912	4·261	0	0
			4·212	1·5 ± 0·5	1
92	234	0·907	4·862	0	0
			4·806	0·0 ± 0·2	0
92	233	0·89 ± 0·01	4·907	0	0
			4·82	0·4 ± 0·3	(0)
			4·59	9·8 ± 1·0	3
90	230	0·900	4·75	0	0, 1
			4·69	0·4 ± 0·9	0, 1
			4·57	6·1 ± 1·0	(2), (3)
			4·49	−3·5 ± 1·1	(0), (0)
90	228	0·889	5·520	0	0
			5·437	3·2 ± 0·4	(2)
			5·433	7·3 ± 0·4	(3)
88	226	0·900	4·881	0	0
			4·694	0·3 ± 0·2	(0)
88	224	0·882	5·784	0	0, 1
			5·547	0·6 ± 0·1	(0), (1)
			5·288	−2·7 ± 0·7	(0), (0)
84	210	0·782	5·401	0	0, 1
			4·600	−1·4 ± 0·3	(0), 0

TABLE VII

Z	A	ρ_0	η_r	${}_1\Delta_r$	j_r
95	241	0·891	5·60	0	⩾5
			5·54	⩽ −18·2	0
94	239	0·901	5·238	0	1
			5·186	5·6±1·1	(3)
			4·83	14·6±2·7	4
92	235	0·91±0·01	4·66	0	5
			4·48	−21·0±1·8	0
			4·28	−23·5±2·7	0
91	231	0·890	5·13	0	5
			5·09	− 7·8±0·7	4
			5·02	− 9·4±0·7	(4)
			4·92	− 7·3±1·2	4
			4·80	−19·9±0·8	1, 0
			4·74	−16·8±1·9	2
91	228	0·87±0·02	6·20	0	4
			5·95	− 6·2±1·8	3
90	229	0·891	5·12	0	5
			5·03	− 7·7±2·5	4
			4·94	−17·8±1·8	2, 1
90	227	0·898	6·159	0	4
			6·129	5·2±0·2	(5)
			6·098	− 1·7±0·1	(4)
			6·078	8·1±0·3	(5)
			6·039	5·0±0·2	(5)
			5·985	− 1·1±0·2	(4)
			5·962	1·7±0·2	(4)
			5·942	2·1±0·3	(4)
			5·924	− 0·5±0·2	(4)
			5·871	−12·7±0·1	1
			5·855	− 5·1±0·2	(3)
			5·826	−13·4±0·1	(1)
			5·773	− 8·7±0·2	(2)
88	223	0·883	5·823	0	3
			5·709	− 4·4±0·1	2
			5·692	4·8±0·3	(4)
			5·634	− 5·2±0·1	(2)
			5·519	− 7·4±0·1	1
86	219	0·885	6·951	0	3
			6·67	− 3·3±1·9	2
			6·554	− 6·6±1·8	1
			6·34	−10·3±1·2	(0)
84	211†	0·79±0·02	7·578	0	4
			7·03±0·04	2·3±1·2	(4)
			6·70±0·04	− 6·0±1·4	3
			6·46±0·06	− 7·2±2·1	3, 2
83	214	0·688±0·025	5·610.	0	2
			5·548	− 3·0±0·1	1
83	212	0·700±0·020	6·198	0	2
			6·158	− 4·4	0
			5·870	− 2·3	(1)
			5·725	0·4±1·1	2
			5·706	− 6·7±0·8	(0)
			5·569	3·2±2·3	3
83	211	0·674±0·020	6·747	0	2
			6·394	− 5·1±0·1	(0)

† This entry refers to the α-disintegration of the isomer ${}^{211}_{84}\mathrm{AcC}'$ (see footnote, p. 35).

disintegration is very probably between states of equal angular momentum, Table VII to cases in which the ground-to-ground state disintegration involves a change of spin. For the results in Table VI, therefore, the first row of entries in Table V should provide sufficient basis of interpretation, for the other results the entries in the other rows of Table V will be required. Limits of error in Δ have been calculated from limits quoted for η and λ in the literature (where such limits are quoted), otherwise they have been assigned somewhat arbitrarily. Almost certainly errors in Δ have been under-estimated in the tables—but without a knowledge of these errors it is impossible to judge the success of any interpretative scheme, and to accept low estimates of error is to impose the more severe test on such a scheme. The bracketing of entries in the sixth columns of Tables VI and VII indicates the degree of success attending our interpretation according to (2.14). For this purpose the constant C has been taken equal to $0{\cdot}71 \pm 0{\cdot}01$. Obviously to have assumed a value much closer to the Gamow value $C = 0{\cdot}372$ would have resulted in an interpretation inherently more improbable, just because larger values of j would have been required ($j \leqslant 7$). Where a j value in the last column either of Table VI or of Table VII is unbracketed the value of ${}_1\Delta_r$ agrees with the 'theoretical' value $C(j_r - j_1)(j_r + j_1 + 1)$ within the combined limits of error of the two values; bracketing indicates that the experimental and theoretical values do not agree within these limits. For three α-emitters represented in Table VI alternative sets of j values are given, the second in each case corresponding to $j_1 = 1$. In these instances the experimental results are hardly accurate enough to decide against unit change in angular momentum in the ground-to-ground state disintegration in each case, and the general concordance between experimental and theoretical values appears to be improved by this assumption.

Here it is appropriate to explain how our empirical value

$$C = 0{\cdot}71 \pm 0{\cdot}01$$

was derived. Table VII was examined to find the maximum range in Δ values amongst the various α-emitters there represented. Maximum ranges of Δ of $23{\cdot}5 \pm 2{\cdot}7$, $19{\cdot}9 \pm 0{\cdot}8$ and $21{\cdot}5 \pm 0{\cdot}3$ were found for three separate species. These were provisionally taken to correspond with ranges in j which were the same in all three cases. Taking count of the information provided by Table V it had to be decided what this range in j was, and it was concluded that the smallest admissible range was $0 \leqslant j \leqslant 5$. This being accepted, then the result $21{\cdot}3 \pm 0{\cdot}4 = 30C$ followed directly. Any other choice (even the choice of a wider range of operative j values) would in fact have led to less satisfactory general agreement between prediction and observation.

The next point of explanation concerns the choice of values of j_1 for the species represented in Table VII. In each case this choice was made to give the best general fit with the experimental values of ${}_1\Delta_r$ given in the table. At that stage the choice was uninfluenced by considerations of the absolute value of λ_1 for each species. It is an important aspect of success of our general treatment that the positions on the Geiger-Nuttall diagrams of the representative points for the ground-to-ground state disintegrations of the α-emitters concerned are fully corroborative of these assignments. In fact, if we had started our analysis by the determination of j_1 for each species on the basis of the absolute value of the partial disintegration constant for the ground-to-ground state disintegration, we should have been led (using our accepted value $C = 0{\cdot}71$) to the same value of j_1 for that disintegration as we actually obtained from the other, entirely independent, method of estimation. This aspect of success appears more significant than the occurrence of a rather large proportion of bracketed j values in column six of Table VII. Particularly in the case of ${}^{227}_{90}$Rd Ac, experimental results are in some considerable doubt in respect of matters of detail.† Again, perhaps more significant than the fact that we have been able to 'explain' all cases of α-particle fine structure on the basis of changes of angular momentum of never more than 5 units as between the ground states of parent and daughter nuclei, is the result that our interpretation involves differences of angular momentum between the various excited states of the daughter nuclei which, for any given nucleus, are very rarely such that de-excitation processes of long lifetime are predicted. The only

† The 'doubtful' groups reported in the last published work on ${}^{227}_{90}$Rd Ac and ${}^{223}_{88}$AcX (Rosenblum, Guillot and Perey, 1937) have not been included in Table VII, but an allowance of about 5 % of the total intensity has been made in each case to take count of such unconfirmed groups.

obvious cases of such a prediction (angular momentum difference greater than three units) occur for $^{235}_{92}U$ and $^{241}_{95}Am$.†

In respect of the differences of spin between the ground states of successive nuclei, one interesting feature of Table VII is that it provides information concerning five α-bodies of the actinium series, besides $^{235}_{92}U$. Spin changes between ground states are given as 5, 4, 3, 3 and 2 units for the disintegrations of Pa, Rd Ac, Ac X, An and Ac C, respectively. The second, third and fourth of these bodies are successive members of the series, and the point which may be made is that it is very much more likely that the relatively large changes of 4, 3 and 3 quantum units of angular momentum which appear to characterize their disintegrations are of alternate sign, rather than that they are all of the same sign, and thus cumulative. Amongst the stable species no case is known of a nucleus with a total angular momentum quantum number greater than $\frac{9}{2}$ (though $I \geqslant 7$ has been reported‡ for the β-active $^{176}_{71}Lu$ (Heyden and Wefelmeier, 1938)). Also it is known that, for $^{231}_{91}Pa$, $I=\frac{3}{2}$, and for $^{207}_{82}Pb$, the end-product of the whole series, $I=\frac{1}{2}$. The following scheme gives I values, for all the nuclei in the series, which are consistent with the conclusions just drawn from the evidence from α-disintegration, as also with the information relative to β-disintegration (see §3.1).

$$\mathrm{Pa}(\tfrac{3}{2}) \xrightarrow{\alpha} \mathrm{Ac}(\tfrac{13}{2}) \xrightarrow{\beta} \mathrm{Rd\,Ac}(\tfrac{11}{2} \text{ or } \tfrac{13}{2}) \xrightarrow{\alpha} \mathrm{Ac\,X}(\tfrac{3}{2} \text{ or } \tfrac{5}{2}) \xrightarrow{\alpha}$$
$$\mathrm{An}(\tfrac{9}{2} \text{ or } \tfrac{11}{2}) \xrightarrow{\alpha} \mathrm{Ac\,A}(\tfrac{3}{2} \text{ or } \tfrac{5}{2}) \xrightarrow{\alpha} \mathrm{Ac\,B}(\tfrac{5}{2} \text{ or } \tfrac{3}{2}) \xrightarrow{\beta} \mathrm{Ac\,C}(\tfrac{5}{2}) \xrightarrow{\alpha}$$
$$\mathrm{Ac\,C''}(\tfrac{1}{2}) \xrightarrow{\beta} {}^{207}_{82}\mathrm{Pb}(\tfrac{1}{2}).$$

We shall see presently that there is further evidence which appears to support our assumption of alternating increase and decrease in I (in α-disintegration) from Pa to An. For this further evidence reference must be made to fig. 12, and to the next section (p. 65).§

† The latest work on $^{241}_{95}Am$ (Asaro, Reynolds and Perlman, 1951) shows six α-particle groups, three of which, of combined intensity less than 0·01 per disintegration, have energy greater than that of the main group ($\eta = 5{\cdot}567$). One can only conclude that the γ-radiation associated with this α-emitter has very unusual characteristics.

‡ The experimental data in this case have recently been reviewed by Klinkenberg (1951), who concludes that $I = 10 \pm 1$.

§ Tomkins, Fred and Meggers (1951) have recently determined the spin of $^{227}_{89}Ac$ as $I=\frac{3}{2}$. If this result is confirmed much that has been written here will require careful reconsideration.

TABLE VIII

Z	A	ρ_0	Δ	j_1
92	228	0·89 ± 0·02	4·8 ± 0·6	2
91	227	0·87 ± 0·02	5·0 ± 0·7	2
90	225	0·871	6·8 ± 0·3	(3)
89	227	0·89 ± 0·01	4·7 ± 0·3	2
89	225	0·88 ± 0·01	5·2 ± 0·6	(2)
89	223	0·87 ± 0·01	7·0 ± 0·7	(3)
88	221	0·880	7·1 ± 1·0	3
87	221	0·88 ± 0·01	3·0 ± 0·4	(2)
87	220	0·87 ± 0·01	7·8 ± 1·0	3
84	213	0·835	1·7 ± 0·7	1
84	211	0·79 ± 0·02	22·5 ± 2·0	5

Hitherto nothing has been said systematically regarding possible spin changes in those α-disintegrations which appear as unique modes of disintegration. It has been stated generally that all the representative points for α-emitters having even A, and $Z = 84$, fall very closely on the smooth Geiger-Nuttall curve of fig. 2, and it has been implied that in each of these, and in many other cases of single-mode α-disintegration, the change of angular momentum as between the states of parent and daughter nucleus involved is zero (or at most one quantum unit). But there are other single-mode α-emitters for which this is obviously not true—if our general scheme of interpretation is accepted. Table VIII treats the best-established cases of this type as the fine-structure disintegrations have been treated in Tables VI and VII. If λ_1 is the experimental disintegration constant, and λ_0 the value obtained from the smooth curve in any case (that is the value corresponding to the experimentally determined disintegration energy η_1), Δ in Table VIII is $(Z-2)^{\frac{1}{2}}\rho_0^{\frac{3}{2}}\log_{10}\left(\frac{\lambda_0}{\lambda_1}\right)$. The entries in column five of the table have been derived using $C = 0{\cdot}71 \pm 0{\cdot}01$ in (2.14), as before, and the figures in the first row of Table V as possible values of Δ/C. The convention previously adopted as regards the bracketing of j values (p. 58) has been re-applied. If the proportion of bracketed values is rather high, it should be emphasized that the use of the smooth Geiger-Nuttall curves as exact reference curves is more arbitrary than the use of the ground-to-ground state disintegration of each species as reference mode with which to compare the other modes of α-disintegration of that species. Values of j_1 in Table VIII should be taken, therefore, merely as indications of the true values—but they are satisfactory in this that, except for the 25 sec. $^{211}_{84}Po$ (Spiess, 1951; Feather, 1952*c*), no value greater than $j_1=3$ is included. If the ground-to-ground state disintegration were to involve spin change of more than this amount in any case, it is very probable that a lower-energy disintegration mode, for which the spin change was less, would compete seriously with the other, and α-particle fine structure would be observed. By contrast, the basis of selection of cases for inclusion in Table VIII was just that no fine structure was observed.

Before leaving the question of spin changes a last remark should be made regarding the implications of the use of a constant value of ρ_0 for all the disintegration modes of a given α-emitter. These different modes arise from the different possibilities in relation to the final state (belonging to the daughter nucleus); the initial state of the transition is the same for all modes—it is the ground state of the parent nucleus. On the basis of the one-body treatment, the radius ρ_0 is assumed to characterize the daughter nucleus, or rather the final state of the transition: there would appear no obvious justification, then, on this basis, for our assumption of constant ρ_0 for all modes. Justification must be sought rather in the results we have already obtained regarding the absolute values of ρ_0 for ground-to-ground state transitions of zero spin change. These results have shown that the variation of ρ_0 with Z for the sequence of representative isotopes from $Z=83$ to $Z=96$, and the variation of ρ_0 with A for Z constant for all these elements, is in each case much more marked than can be accounted for on a purely geometrical interpretation (this point will be taken up again later, see p. 67). We explain these considerable variations as reflecting the varying degrees of 'difficulty of formation' of the nascent α-particle in the parent nucleus (see Nordström, 1951). This would appear to be a feature common to all modes when the fine-structure phenomenon is in evidence. In adopting for a given α-emitter the same value of ρ_0 for all fine-structure modes, we are thus assuming that this 'difficulty of formation' effect is the major effect causing departure of ρ_0 from the geometrical value. It should be noted that the nature of the effect is that initial formation of the α-particle becomes progressively more difficult as Z decreases from 90 to 84

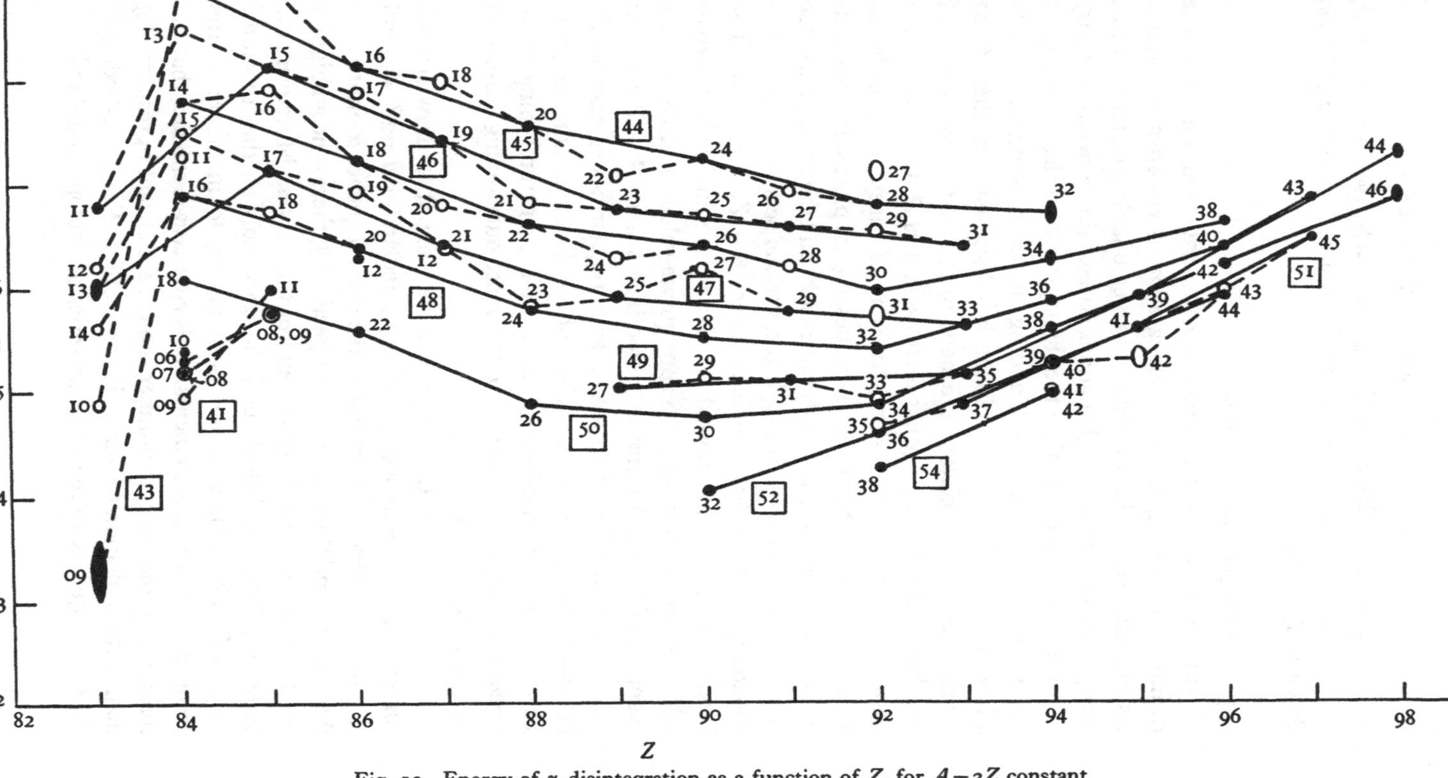

Fig. 12. Energy of α-disintegration as a function of Z, for $A-2Z$ constant.
η in MeV., ○ odd N, ● even N.

—and markedly so between $Z=84$ and $Z=83$ (see fig. 10)—and, for Z constant, progressively more difficult as N decreases, at least down to $N=126$ (see fig. 11).

2.4. Disintegration energies

Until now we have accepted the experimental values of the disintegration energies in α-disintegration as empirically 'given'. As we have seen, the theorist is as yet unable to predict these energies as the result of calculations based upon a detailed theory of nuclear structure. On the other hand, it is clear that the energies are not randomly distributed, and fig. 12 has been drawn so as to allow the regularities which obviously subsist amongst them to be more directly appreciated. We have in fact noted one aspect of regularity already: the regular increase in η as A decreases for the members of an isotopic group (that is, for Z constant). A different aspect is stressed in fig. 12. Here values of η are plotted (for ground-to-ground state transitions only) against Z, and the attempt is made to draw smooth curves through the various sets of points corresponding to which the isotopic number, $A-2Z$, is constant. The 'points' themselves are plotted as full or open circles as N is even or odd, and are labelled with the appropriate values of A (A–200 for economy of detail). In drawing smooth curves for $A-2Z$ constant we are concentrating attention on such regularities as may obtain in relation to the successive values of the disintegration energy in a sequence of pure α-disintegrations. This is a natural thing to do, since these sequences constitute the most outstanding feature of similarity between the series.

It is obvious from the figure that the attempt to draw smooth curves is again broadly successful; in particular, for $A=2Z=50$ and $A-2Z=48$ (values appropriate to the main sequence α-emitters of the uranium and thorium series, respectively), there is no ambiguity at all concerning the detailed trend of the curves. Moreover, the points for ${}^{218}_{85}\mathrm{At}$ and ${}^{214}_{83}\mathrm{Ra\,C}$ both lie reasonably on the 'thorium' curve ($A-2Z=48$), although both belong to members of the uranium series of radioelements. This emphasizes the fact that the constancy of the isotopic number is more significant in respect of η than is the affiliation to one series or another. For the α-bodies of the main sequence of the actinium series the isotopic number is 47, and six α-emitters of the neptunium $(4n+1)$ series also have this value of $A-2Z$. Again, there is no systematic separation of the points in a way which would suggest two curves, one for each series of elements, but it is to be noted that the precise trend of the common curve is less well defined by the points than is the case for $A-2Z=50$ or 48.†

This feature of fig. 12 merits some discussion. Purely empirically it would appear that η is greater for the α-disintegrations of $^{227}_{90}$Rd Ac and $^{219}_{86}$An of maximum energy, and less for the α-disintegration of maximum energy observed with $^{223}_{88}$AcX, than would be expected on the basis of the general regularities exhibited by the curves for $A-2Z=48$ and 50. We are tempted to ascribe these clear regularities (over the range $84 \leqslant Z \leqslant 92$) to the fact that for $A-2Z=50$ and 48 we are dealing throughout with ground-to-ground state transitions occurring without change of spin (or, in certain cases with change of spin no greater than one quantum unit)—and to trace the irregularities for $A-2Z=47$ to the fact that relatively large spin changes are involved. If we do this, the alternating sign of the discrepancy exhibited by the points for RdAc, AcX and An naturally suggests an alternation in the sign of the spin change—which is our previous suggestion (the point for $^{215}_{84}$AcA, for which $j_1=1$ or 0, lies on the curve). Following this correlation further, we note that we are associating excess energy set free in disintegration with decrease in nuclear spin in the disintegration process—and a deficit of energy, as with AcX, with an increase in the total angular momentum as between initial and final nuclei. This is a reasonable result: it has been shown (Guggenheimer, 1942) that if the angular momentum of a nucleus is due to the rotation of the nucleus as a whole (or to some considerable part of the nucleus with respect to the rest), then, when the angular momentum decreases in a disintegration process, practically the whole of the kinetic energy corresponding to the change of spin is available for liberation in the disintegration. Passing from the qualitative to the quantitative aspect of the matter, if we are 'explaining', for example, the excess disintegration energy of about 3×10^5 eV., which appears in the

† If attention is confined to the full-circle points (N even) then the curve for isotopic number 47 on fig. 12 is just as regular as are the curves for $A-2Z=50$ and 48.

disintegration $\mathrm{RdAc} \xrightarrow{\alpha} \mathrm{AcX}$, by a decrease of total angular momentum from $\frac{11}{2}$ to $\frac{3}{2}$ quantum units, the fundamental energy $h^2/8\pi^2 J$ (J = moment of inertia concerned) is 10 keV. and the corresponding value of J may be calculated as $3{\cdot}4 \times 10^{-47}$ g.cm.^{2}. It will be observed that this moment of inertia is definitely of the correct order of magnitude: for the rotation of the whole nucleus ($A = 227$) as a rigid sphere it would imply a radius of $4{\cdot}8 \times 10^{-13}$ cm., whilst a single neutron rotating in a circular orbit of radius $4{\cdot}6 \times 10^{-12}$ cm. would have the same moment of inertia about the axis of the orbit. Certainly the true radius of the radioactive nucleus lies between the values deduced on the basis of these limiting assumptions.

Our discussion of fig. 12 has so far centred on the sequences $A-2Z = 47$, 48 and 50. Of these the last contains species of even neutron number only, and the next-to-last only two species of odd neutron number ($^{218}_{85}\mathrm{At}$ and $^{214}_{83}\mathrm{RaC}$) along with eight of even neutron number. We have already remarked (footnote, p. 65) that if attention is confined to species of even neutron number (full-circle points on fig. 12) all three sequences are represented by quite regular lines on the figure. This result may be generalized with striking success to apply equally to the sequences having isotopic numbers 46, 45 and 44 (each of which contains even- and odd-N species in numbers large enough to be significant)—and, so far as our incomplete information goes, to the sequences $A-2Z = 49$, 51 and 52. We may say, then, that for $Z \geqslant 84$, $N \geqslant 128$, the α-disintegration energies (for ground-to-ground state disintegration) of the $\binom{e}{e}$ species of each even value of isotopic number ($A-2Z \geqslant 44$) and those of the $\binom{e}{o}$ species of each odd isotopic number ($A-2Z \geqslant 45$) vary in each case ($A-2Z$ constant) smoothly with Z. And, equally important, we may say that similar statements cannot be made regarding the α-disintegration energies of the $\binom{o}{o}$ species of constant (and even) isotopic number and the $\binom{o}{e}$ species of constant (and odd) isotopic number. In each of these cases the variation of disintegration energy with Z is more or less irregular.

In the range not here considered, between $Z = 84$ and $Z = 83$, the four curves for isotopic numbers 45, 46, 47 and 48 on fig. 12 show very similar reversals of trend: disintegration energies decrease by amounts of the order of 1·5 MeV. between these values of Z. Much larger decreases, of the order of 4·0 MeV., are recorded by the more abrupt discontinuities exhibited by the curves for $A-2Z=43$ and 44. The distinction here is that for $Z=83$, $A-2Z=45$, 46, 47 or 48, $N \geqslant 128$, whilst for $A-2Z=44$, $N=127$ and for $A-2Z=43$, $N=126$ for this value of Z. It would appear that there is a decrease in α-disintegration energy when the atomic number of the parent species decreases from 84 to 83 in all cases, and a superadded decrease when the neutron number decreases from 128 to 127 or from 127 to 126 (Glueckauf, 1948). It is probable that the curve for $A-2Z=42$ will show a decrease in η between $Z=85$ and $Z=84$ which, being smaller in magnitude than those last considered (of the order of 1·8 MeV.), may be attributed solely to the decrease in N from 127 to 126 (set against any small rise attributable to the decrease in Z from 85 to 84). In this general way the crossing of the curves of fig. 12 in the region $83 \leqslant Z \leqslant 85$ may be systematized. There is some evidence for similar, though much less marked crossings in the region $Z>92$, but the evidence is not as yet sufficiently clear-cut for an attempt at analogous systematization to be profitable.

The regularities which we have just noted have been exhibited in a different, and striking, form by Pryce (1950). Instead of the experimental disintegration energy η, Pryce has plotted $\eta-\eta_c$ against A. η_c is the semi-theoretical disintegration energy calculated from the v. Weizsäcker formula

$$\left.\begin{aligned} \eta_c &= Q(A,Z)-Q(A-4,Z-2)-\eta_0, \\ Q(A,Z) &= \beta\frac{(A-2Z)^2}{A}+\gamma A^{\frac{2}{3}}+\delta\frac{Z^2}{A^{\frac{1}{3}}}. \end{aligned}\right\} \qquad (2.15)$$

It should be noted that this formula is based on the assumption that the nuclear volume is directly proportional to A, and the magnitude of η_c is indeed most sensitive to this assumption. In the case of a typical heavy α-emitter, the contributions to η_c corresponding to the second and third terms in the expression for $Q(A,Z)$ are of the order of 7 and 30 MeV., respectively (these are the contributions from the changes in surface energy and coulomb energy as the result of α-emission), and, if the radius of the final nucleus were to be 1 % smaller than corresponds to the regular $A^{\frac{1}{3}}$ variation, the formula predicts that a further 4 MeV. of energy would be available from the contraction alone (β assumed constant).

Pryce's representation is shown in fig. 13,† only the smooth curves being given, the points being omitted. It will be observed, first, that the whole range of values of $\eta-\eta_c$, for species having

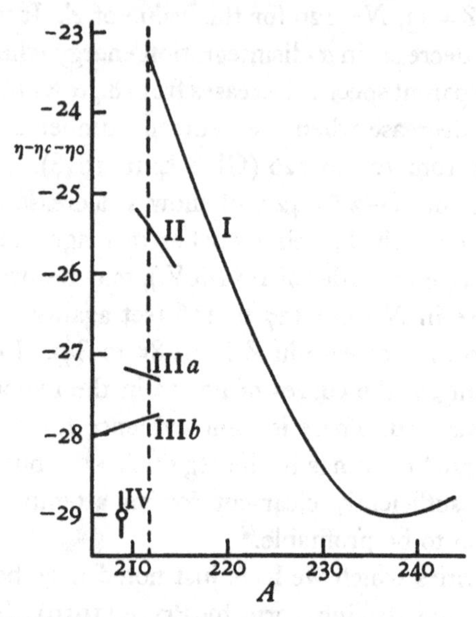

Fig. 13. Difference between observed and 'calculated' values of α-disintegration energy as a function of A. $\eta-\eta_c-\eta_0$ in MeV.

$205 < A < 245$, is less than 6 MeV., and, secondly, that this quantity appears to be purely A-dependent, once the existence of three separate branches to the curve of variation is accepted. On the main branch the fit of individual points is generally within about 0·15 MeV.

That the displacement and branching of the curve about $A=210$ to 214 is significant is obvious from the figure; its correlation with structural pattern can be expressed as follows. Points falling on the main branch (I) are those of α-emitters for which $N \geqslant 128$, $Z \geqslant 84$,

† Here $\eta-\eta_c-\eta_0$ is plotted, rather than $\eta-\eta_c$, since in (2.15) the greatest uncertainty is in the appropriate value of η_0. We may take, approximately, $\eta_0=29\pm 1{\cdot}5$ MeV.

points falling on the second branch (II) belong to α-emitters with $N=127$, $Z \geqslant 84$ and $N \geqslant 128$, $Z=83$, points falling on the third branch (III) are those of α-emitters with $N \leqslant 126$, $Z \geqslant 84$ and of RaE ($N=127$, $Z=83$).† Remembering that the structural units forming the α-particle are two neutrons and two protons, this statement may be further simplified. The energy available for α-disintegration (in excess of that calculated according to (2.15)) appears to depend upon the number of nucleons available for α-formation outside the 'closed' systems of 126 neutrons and 82 protons which constitute the nucleus $^{208}_{82}$Pb. The excess energy is greatest when the number of nucleons so available is precisely four (two neutrons and two protons), and with an increase in this number it decreases smoothly to a minimum when there are 24–26 nucleons available. When one nucleon (neutron or proton) has to be supplied from within a closed system the excess is in general less—and again it decreases with increasing numbers of 'outside' nucleons. When two nucleons have to be supplied from the closed systems, to match two from outside, the excess is further reduced, but it does not, according to present scanty information, appear to vary very much with the number of 'outside' nucleons available. Finally, if our estimate of the disintegration energy of $^{209}_{83}$Bi is correct (see fig. 12)‡ the point for this species on fig. 13 lies at about $\eta - \eta_c - \eta_0 = -29{\cdot}0$, and a fourth and still lower curve for species in which only one nucleon is available outside the closed systems appears to be indicated. It is probable that a fifth (and lowest) curve completes the diagram to refer to those species for which all the constituents of the emitted α-particles have to be provided from within the system, that is to parent species for which $N \leqslant 126$, $Z \leqslant 82$ (see Feather, 1952*b*).

Purely formally, the results exhibited on fig. 13 might be interpreted in terms of a change of nuclear volume on α-disintegration slightly different from that predicted on the simple assumption of constant density. Such a change would be cumulative for the

† From the diagram this branch appears significantly double, the doubling corresponding to the distinction between species for which $N=126$ (III*a*) and those for which $N<126$ (III*b*), when in each case $Z \geqslant 84$. The point for $N=127$, $Z=83$ appears to lie on III*a*.

‡ Fig. 12 was drawn before the observations of Faraggi and Berthelot (1951) were made (see footnote, p. 2).

successive α-emitters of a disintegration series, but it would obviously be stretching the interpretation too far to attempt a correlation with the trend of ρ_0 exhibited in fig. 10. The results of figs. 10–13 all point to a discontinuity in nuclear structure beyond $^{208}_{82}Pb$, ($N=126$, $Z=82$) but they do so in different ways.

2.5. Discontinuities in nuclear building and the α-activity of lighter nuclei

In this section we attempt to discover, for smaller values of N and Z, discontinuities in nuclear building which might issue in α-activity. We have just seen that, in comparison with the isotopes of element 84, the isotopes of element 83 are inconspicuous as α-emitters for two distinct reasons—on account of the smallness of the energy available, and because of an intrinsic lowering of disintegration probability. On the other hand, as Z decreases from 90 to 84 α-activity becomes progressively more and more pronounced: there is thus a real maximum of instability at $Z=84$, at least over a considerable range of A. We have now to inquire whether a similar phenomenon occurs in any other range of Z. We have already concluded that the magnitude of the phenomenon at $Z=84$ is heightened by the occurrence of an overlying discontinuity at $N=128$, affecting the same species as are affected by the discontinuity in Z (see Wapstra, 1951), but we may still look for a phenomenon, similar in nature but on a reduced scale, in cases where the structural pattern is discontinuous in relation to N or Z alone. In just such circumstances α-emission may take place for species which obviously do not belong to one or other of the classical series of radioelements.

In an attempt to locate these discontinuities we refer again to fig. 1. Two dotted lines have been drawn across this diagram defined by $A-3Z=-30$ and $A-3Z=-38$. These we shall take as representing the general trend of the lower stability limits for species within certain ranges of Z which are clear from inspection of the figure. On this basis we may say that obvious discontinuities occur in the regions about $Z=40$ and $Z=60$. We have already noticed one aspect of this result in our earlier classification of $^{88}_{40}Zr$ and $^{140}_{60}Nd$ as 'missing' stable species (p. 22). From our present standpoint it is interesting to observe that the nature of the discontinuities revealed by this interpretation of fig. 1 is that between $Z=38$ and $Z=42$ ($N \sim 50$), and likewise between $Z=58$ and $Z=62$ ($N \sim 82$), it seems as though some change in structure occurs whereby stability may be preserved even though the specific charge of the nucleus (or the ratio Z/A) becomes greater than the previous general trend of the lower stability limit would lead us to predict. Assuming a spherically symmetrical nucleus we might correlate this change with an abrupt increase in radius, but equally well some other basis of interpretation might eventually prove to be correct (see Wapstra, 1952). However, the situation is sufficiently analogous to that which we have encountered among the heavy radio-elements to make it profitable to look for hitherto undetected α-activity in elements for which $Z \sim 40$ and $Z \sim 60$, respectively. As is well known, the only established case of 'natural' α-activity in an element for which $Z<83$ occurs with $^{147}_{62}$Sm (Weaver, 1950; Rasmussen, Reynolds, Thompson and Ghiorso, 1950).† To this extent our speculations would appear to be the more securely founded. Also, we have seen (p. 23) that in order to import regularity into the sequence of nuclei of odd Z (which do not enter into the definition of the lower stability limits represented by the dotted lines in fig. 1) we have found it necessary to assume precisely the same type of 'expansive' discontinuity between $Z=57$ and $Z=59$. In that case we found it useful to regard $^{139}_{57}$La as ending one sequence, and $^{141}_{59}$Pr as starting another sequence, of series of odd isotopic number. For both these species, it should be noted, $N=82$.

The case of samarium will repay further consideration before we turn to a discussion of the possible α-activity of other elements. The isotopic constitution of this element is unique in that no other example is known in which a difference of four units exists between the mass numbers of the lightest and next-lightest even isotopes. $^{144}_{62}$Sm is known (as also $^{148}_{62}$Sm), and $^{147}_{62}$Sm is present in ordinary samarium and is α-active. But $^{146}_{62}$Sm has not been detected. This constitution would receive reasonable explanation if it were assumed that for $Z=62$ the maximum α-instability occurs for $A=146$

† Actually the isotopic assignment of this activity has been variously reported in the past. It would appear, however, that the present assignment is incontrovertible.

($N=84$). To assume this is clearly to assume that we have here a discontinuity which is N-determined—the significant number being $N=84$, rather than $Z=62$.

If $^{146}_{62}Sm$ were to prove to be the most unstable of the α-emitting isotopes of samarium, we should expect $^{145}_{61}Pm$ to be the most α-active of the isotopes of the 'missing' element 61. We have already seen that $^{145}_{61}Pm$ can, not unreasonably, be regarded as stable both against β-emission and capture transformation (p. 23),† and we were led to postulate that the near-absence of element 61 on the earth is due to the α-instability of this isotope. We are now able to recognize further support for this assumption. Moreover, there is already some experimental evidence (see p. 75) for the attribution to element 61 of an α-activity of rare-earth origin (α-particle range 1·8 cm. in standard air), and it may well be that $^{145}_{61}Pm$ is the responsible species. If $^{145}_{61}Pm$ is the responsible species, and is capture-stable, then, as already pointed out (p. 17), $^{145}_{60}Nd$ must be β-active. We shall see later (p. 138) that if the reported β-activity of neodymium is a reality the active isotope must almost certainly be of odd mass number.

As regards the possible α-activity of neodymium, extension of the series $N=84$ brings us to $^{144}_{60}Nd$. The experimental evidence is that here any energetically possible activity has sunk well below the limits of detection. As a corollary we conclude further that $^{140}_{60}Nd$ is not missing in nature because of α-activity. The anomaly is rather that $^{144}_{62}Sm$ is a 'supernumary' species (p. 26)—and the series of stable nuclei of isotopic number 20 would have ended with $^{136}_{58}Ce$ had not $^{144}_{62}Sm$ been favoured because of the specially strong binding of the system of 82 neutrons which it contains. (Similarly, the first member of the series of isotopic number 28 would have been $^{148}_{60}Nd$, not $^{136}_{54}Xe$, but for a like reason.)

The significance of the neutron number 84 for the α-activity of the rare earth elements has recently been further emphasized by investigations on the neutron-deficient isotopes of these elements. Thompson, Ghiorso, Rasmussen and Seaborg (1949) first observed that bombardment of dysprosium with high energy protons produced three α-active bodies (periods of the order of minutes or hours; disintegration energies of the order of 4 MeV.), and that

† See, however, footnote, p. 17.

two of these bodies were also produced by similar bombardment of gadolinium. Now Rasmussen, Reynolds, Thompson and Ghiorso (1950) have established by the method of mass analysis that the 4·0 hr. period (disintegration energy 4·1 MeV.) is due to $^{149}_{65}$Tb. It is not improbable that the other periods are to be attributed to neighbouring species also having $N=84$ (or 85).† We shall recur to the question of the bearing of the $N=84/82$ discontinuity on the problem of the emission of α-particles in fission at a later stage (p. 76), meanwhile we may remark that for N constant (and in this case equal to 84) there appears to be a general rise in α-disintegration energy from $Z=60$ to $Z=65$ similar to that which is evident from fig. 12 for each value of N from 129 to 147, over the range of Z increasing appropriate to each. We conclude from this that there is no marked Z-determined discontinuity in α-disintegration energy over the range $60 \leqslant Z \leqslant 65$.

If the general evidence from fig. 1 indicates an abrupt structural discontinuity for $Z \sim 60$, it is even more definite in showing an abrupt change in a similar sense for $Z \sim 40$. Here the character of the 'missing' species $^{88}_{40}$Zr and $^{90}_{38}$Sr is most obviously in question. It is just possible that 'unexplained' helium which has been found in zircons of great geological age (Strutt, 1910) may be held to afford some evidence supporting the possibility that $^{88}_{40}$Zr—or some other isotope of zirconium—might be α-active, but we should be careful in accepting this as evidence since zircons quite frequently contain helium, in smaller quantities, which is obviously derived from traces of heavy radioactive 'impurities' in the mineral. On the basis of present ideas $^{90}_{38}$Sr ($N=52$) would appear more likely as an α-emitter than $^{88}_{40}$Zr ($N=48$).

If we pursue the geological evidence, there are two other minerals which sometimes contain quite large amounts of helium the origin of which is uncertain. These minerals are sphene ($CaSiTiO_5$) (Kevil, Larson and Wank, 1944) and beryl (Strutt, 1908). It is a very interesting fact that, in addition to neodymium and zirconium, there is just one other element ($Z>16$) for which the difference between the mass numbers of the two stable isotopes

† An α-emitter of half-value period 7 hr. has recently been reported as produced by α-particle bombardment of samarium (Sun, Pecjak, Jennings, Allen and Nechaj, 1951) and more detailed information regarding the new activities has been given by Rasmussen, Thompson and Ghiorso (1951).

of smallest mass is one unit only. This element is titanium. In the same sense as we may say that $^{140}_{60}$Nd and $^{88}_{40}$Zr are missing stable species, we might conclude, therefore, that $^{44}_{22}$Ti is missing, also. If this conclusion is significant, then the helium in certain sphenes might be attributed to the α-activity of $^{44}_{22}$Ti—so long as the energy of α-emission were known to be sufficiently small (vide infra).

In this connexion it should be noted that in a very real sense $^{44}_{22}$Ti is the exact analogue of $^{212}_{84}$Th C′: in each case there is precisely one α-particle equivalent (two neutrons and two protons) outside a system which is 'closed' in respect both of N and Z. $^{44}_{22}$Ti is not at present listed as a known capture-active species.

As concerns the occurrence of helium in beryls, expert opinion now appears to be veering round to the view (Fay, Glückauf and Paneth, 1938; Khlopin and Abidov, 1941) that this is to be attributed to the α-activity of small quantities of an unidentified impurity, which has probably decayed completely during the course of geological time, rather than to any natural or 'artificially' produced activity specific to beryllium itself (either to the hypothetical $^{8}_{4}$Be or to $^{9}_{4}$Be). In this connexion we might hazard the guess that the impurity is likely to be one of the elements involved in the discontinuities at $Z \sim 60$, $Z \sim 40$ (or even, possibly, $Z \sim 20$).

In discussing the origin of the helium in sphene we have made the proviso that, for our suggestion to be admissible, the energy of α-emission of $^{44}_{22}$Ti must be sufficiently small. This proviso is necessary to ensure that the lifetime of the supposed α-emitter is long enough for the survival of the active species beyond the earlier stages of terrestrial evolution. In order to gain a clearer estimate of what is required, Table IX has been compiled. Half-value periods are given, for various energies of disintegration, for each of the species $^{136}_{52}$Te, $^{90}_{38}$Sr and $^{44}_{22}$Ti, calculated on the basis of (2.7) and (2.8) and assuming $\rho_0 = 0{\cdot}14A^{\frac{1}{3}}$ (see fig. 10). It may be noted that these calculated values are not inconsistent with the accepted lifetime of $^{147}_{62}$Sm ($1{\cdot}1 \times 10^{11}$ years, for $\eta = 2{\cdot}24$),† and though extrapolation over such a wide range is admittedly dangerous it may be assumed that they indicate the general variation with sufficient accuracy. Our immediate conclusion from the table is that for the helium in sphene to be due to the α-disintegration of $^{44}_{22}$Ti the

† 'Calculated' lifetime 4×10^{12} years.

disintegration energy would certainly have to be less than 0·5 MeV. This is not, however, an impossible condition (see Ballou, 1949), though the undetectably low abundance of $^{44}_{22}$Ti at the present day presents certain difficulties for this interpretation.

TABLE IX

Species \ η	0·36	1·0	2·0	3·0	9·0
$^{136}_{52}$Te	—	—	$2·5 \times 10^{7}$ y.	30 m.	4×10^{-18} s.
$^{90}_{38}$Sr	—	$2·5 \times 10^{12}$ y.	7 m.	2×10^{-6} s.	$3·5 \times 10^{-21}$ s.
$^{44}_{22}$Ti	3×10^{13} y.	0·1 s.	3×10^{-11} s.	—	—

We may draw another conclusion from Table IX, this time concerning our earlier suggestion that the α-particles of 1·8 cm. range, of uncertain origin, might in fact be emitted in the disintegration of $^{145}_{61}$Pm. The value of η corresponding to this range is 3·2, and since Table IX indicates that for $^{136}_{52}$Te and $\eta = 3·0$ the half-value period is 30 min., it is at once obvious that $^{145}_{61}$Pm cannot, by many powers of ten, be sufficiently long-lived in relation to 1·8 cm. range α-particle emission† to provide any explanation of the facts—unless this α-active species be the daughter product of a much longer-lived parent. For the facts are (it is assumed that they refer to the same group of α-particles) that α-particles of 1·8 cm. range are emitted by some element with rare-earth-like properties, which exists in small amount on the earth at the present time (Schintlmeister, 1936) and that certain pleochroic haloes provide evidence both for α-particles of this range and also for a group of particles of about 1·2 cm. range, probably due to samarium (Henderson and Turnbull, 1934). The evidence from pleochroic haloes, of course, refers to α-activity continued over long periods of the past. To explain these facts, very obviously a controlling period of the order of 10^{9} years, at least, is required.‡ Now in this case it so happens that the suggestion of a long-lived parent is ready to hand. We have already made the suggestion that $^{145}_{60}$Nd is β-active and of long life.§ The attribution of the α-particle

† More exact calculation gives a half-value period of 2·7 years for this species.

‡ It remains to be seen whether the report of the emission of α-particles of 1·8 cm. range by ordinary bismuth (cf. footnote, p. 2) is relevant to the present discussion.

§ See, however, footnote, p. 17.

group of 1·8 cm. range to $^{145}_{61}Pm$ merely gives definiteness to our previous suggestions: these may now be represented by the following tentative scheme

$$^{145}_{60}Nd \xrightarrow{\beta} {}^{145}_{61}Pm \xrightarrow{\alpha} {}^{141}_{59}Pr.$$

The reason for the inclusion of very short half-value periods ($< 10^{-14}$ sec.) in Table IX is that we may discuss briefly the rare emission of α-particles (of energies from the smallest detectable to about 26 MeV. (Allen and Dewan, 1950)) as 'secondary' particles in fission. These particles, about 80 % of which have energies less than some 2 MeV.† (Cassels, Dainty, Feather and Green, 1947; Green and Livesey, 1948), appear to be emitted within about 10^{-13} sec. of the instant of fission in all cases. First of all it is clear that we cannot explain the rapid emission of the particles of smallest energy (say, less than 2·5 MeV.) in terms of the normal α-disintegration of any nucleus of charge number between 35 and 60 (roughly the range of Z covered by the primary products of fission). But to explain the rarity of the emission of secondary α-particles, either α-particles of small energy from the initially much distorted—or particles of any energy from the otherwise highly excited (Kinsey, Hanna and van Patter, 1948; Deutsch and Rotblat, 1951)—nuclei of the primary fission products, would appear to require the assumption that intrinsically only a few of these product nuclei in their normal states have an appreciable tendency towards α-instability. That is precisely the conclusion towards which the whole of our present discussion has been moving: this tendency seems to be confined effectively to nuclei for which $N = 84$ (or 85) and N (or Z) $= 52$ (or 53).

Fulfilling these conditions are the primary fission products $^{87}_{35}Br$, $^{88}_{35}Br$, $^{131}_{52}Te$, $^{134}_{52}Te$, $^{135}_{52}Te$, $^{137}_{53}I$ and $^{138}_{53}I$, all produced in fairly large yield in the slow-neutron fission of $^{235}_{92}U$. The suggestion is, then, that these primary products, whenever they are formed with high energy of excitation, emit α-particles rather than γ-quanta.‡ Here it may be noted that $^{136}_{52}Te$ has not, as yet, been identified as a (primary) product of fission—though its hypothetical daughter $^{136}_{53}I$ is known to

† Evidence that many of these short-range particles have $Z > 2$ appears to be accumulating (see Titterton, 1951).

‡ In order to explain the angular distribution of the 'secondary' particles, this emission must occur within about 10^{-21} sec. of the separation of the fragments.

occur (Stanley and Katcoff, 1949). This species, ${}^{136}_{52}\mathrm{Te}$, with $Z=52$, $N=84$, clearly presents the most favourable case of all for α-emission: it may well be that it is markedly α-active even in its ground state.

In regard to some of the less frequent occurrences in fission the above example shows that the concept of nuclear shell structure is almost certainly significant for interpretation: for the more normal modes attempts have likewise been made to apply this concept (Meitner, 1950), in one case to explain the fine structure in the mass-yield curve. Glendenin and co-workers (Glendenin, Coryell and Edwards, 1949) and Thode and collaborators (Macnamara, Collins and Thode, 1950) have together examined this matter in detail over the mass number range $131 \leqslant A \leqslant 137$, and both groups are agreed that the yield of fission chains having $A=133$ is at least 20% higher, and that for $A=134$ about 35% higher, than would be expected on the basis of the smooth curve of yield against A. For $A=135$ the yield is normal again (Glendenin and Coryell, 1950). The two high yields are obviously correlates of the N values 82 and 83 (Glendenin, 1949): the chain $A=133$ probably derives chiefly from the primary product ${}^{133}_{51}\mathrm{Sb}$ ($N=82$), and it is likely that ${}^{134}_{51}\mathrm{Sb}$ ($N=83$) and ${}^{134}_{52}\mathrm{Te}$ ($N=82$) provide the main contributions to the chain $A=134$ (Pappas, 1951).

As a solution of the general problem of the relative stability of nuclei, direct determination of mass undoubtedly affords the possibility of a final appeal to experiment, but it will probably be a considerable time before determinations are available, for all 'stable' species, which are sufficiently accurate for the merits of any hypothesis of shell structure to be tested adequately on this basis. However, there is already some recent evidence, chiefly from the work of Duckworth and his colleagues, which appears to support the suggestion that the most significant discontinuity in stability amongst the medium-heavy elements occurs around the neutron number 50. Values of the (negative) 'packing fraction', or mass-defect per nucleon (D/A), have been determined (Duckworth, Woodcock and Preston, 1950; Duckworth, Preston and Woodcock, 1950) for certain $\binom{e}{e}$ species in the mid-mass range with the results shown in Table X. The figures require no further comment.†

† For more recent determinations, see Duckworth, Kegley, Olson and Stanford (1951).

Hardly less direct than the evidence from mass determinations—and generally more easy to obtain with the requisite accuracy—is the evidence provided by the measurement of reaction energies, in particular the energies of (d, p) reactions. Harvey (1951) has published the results of extensive studies in this field which show the marked change in strength of binding as between the 126th and 127th, and between the 50th and 51st, neutron in several nuclei.

TABLE X

Species	$^{50}_{24}Cr$	$^{52}_{24}Cr$	$^{56}_{26}Fe$	$^{60}_{28}Ni$	$^{90}_{40}Zr$	$^{96}_{42}Mo$	$^{100}_{42}Mo$
$D/A \times 10^4$	7·96 ±0·05	8·25 ±0·05	8·03 ±0·05	8·60 ±0·05	7·58 ±0·07	6·67 ±0·04	6·14 ±0·04

Passing to the other extreme, evidence of a very indirect character is provided by a study of the isotopic hyperfine structure ('isotope shift') observed in atomic spectra. For the lightest elements this effect is predominantly a pure mass effect (Bohr, 1913), in which the fractional difference in wavelength between corresponding lines of neighbouring isotopic species is of the order of $5 \times 10^{-4} A^{-2}$, but the effect observed with heavier elements, sometimes of magnitude a hundred times larger than that of the pure mass effect, must be ascribed to another origin. Such an effect was first observed with thallium by Schüler and Keyston in 1931, and first interpreted as a nuclear volume effect by Racah (1932) and Breit (1932, 1934) shortly afterwards. It is consonant with this interpretation that in general the isotopic hyperfine structure pattern for any element—at least for the isotopes of even A—consists of a set of roughly equally spaced components, and that the magnitude of the pattern spacing is greatest for those spectral terms which belong to electron states of greatest nuclear penetration. In 1934 Schüler and Schmidt reported an anomaly in the spacings of the pattern for the isotopes of samarium of even A, and in 1945 Klinkenberg detected a similar anomaly in the pattern for neodymium. These observations were confirmed by Brix and Kopfermann in 1948. Since that time a great deal of investigation, both experimental and theoretical (Crawford and Schawlow, 1949; Breit, Arfken and Clendinin, 1950), has been directed towards elucidating the whole phenomenon. Experimentally the most nearly complete and most significant results are those for the $\binom{e}{e}$ isotopes of the elements from tellurium ($Z = 52$)

to gadolinium ($Z=64$). Over this range of Z it appears that any anomalies in the isotope shift are almost entirely N-dependent. Independently of Z, term energy differences for neighbouring isotopes of even $A(\Delta A=2)$ appear to fall into three classes depending upon the change in N involved (Brix and Frank, 1950). If energy differences corresponding to the pairs of neutron numbers 82 and 84, 84 and 86, 86 and 88, and 90 and 92 are regarded as normal—as they appear to be—in relation to the original theory of the effect, those corresponding to the pairs 78 and 80, and 80 and 82 appear small, and, finally, those corresponding to the pair $N_1=88$, $N_2=90$ seem large, by the same standard.† The anomalies first recognized with samarium ($Z=62$) and neodymium ($Z=60$) are the unexpectedly large spacings of this last class corresponding to $A_1=150$, $A_2=152$ in the first case and to $A_1=148$, $A_2=150$ in the second.

In the interpretation of these various anomalies it is useful to bring into the experimental picture the results concerning the isotopes of odd A in the same general range of Z. Usually, for even Z, the term values for isotopes of odd A are 'staggered' with respect to those of even A in the sense of A decreasing—though not in general by amounts great enough to change the natural order of the terms. Thus, the isotopic hyperfine structures of several lines in the violet and ultraviolet spectra of neutral lead, for example (Manning, Anderson and Watson, 1950), are represented by frequency differences having in common the ratios 1·7:0·77:1·22. On the assumption of regular spacing these ratios should be 2:1:1 ($\Delta A=206$–204, 207–206, 208–207, respectively). Breit, Arfken and Clendinin (1950) have introduced the idea of polarization of the nucleus by the atomic electrons as a possible basis of explanation of the apparently greater-than-normal attraction between nucleus and electron (or apparently smaller-than-normal nuclear volume) required to systematize these results for the odd A isotopes of such elements (of even Z). Greater nuclear polarizability is here correlated with the greater density of low-lying excited states which is known to characterize the odd-A nuclei. If evidence from the range $52 \leqslant Z \leqslant 64$ gives some support for an explanation of this type, it also shows how difficult it is to draw clear conclusions,

† A statement of the situation, having a slightly different emphasis, has recently been given by Murakawa and Ross (1951).

in the present state of knowledge, from observations on the isotope shift. Thus, with barium, Arroe (1950) has observed a staggering so great as to throw the isotopic hyperfine structure components far out of the natural order of mass numbers, and of a magnitude which depends on the electron transition concerned (one line in the spectrum of neutral barium shows a staggering in which the 137 mass-number component lies on the low-mass side of that corresponding to $A=134$). All this clearly calls for an explanation over and above that provided by a pure volume effect, but, on the other hand, in the same range of Z, xenon provides an example (Koch and Rasmussen, 1950) of the very rare phenomenon of odd-mass component staggering in the direction of increasing A. This would require less-than-normal polarizability for the xenon nuclei of odd A, yet the evidence from nuclear isomerism shows that, for the three species $^{131}_{54}Xe$, $^{135}_{56}Ba$ and $^{137}_{56}Ba$ which exhibit this effect, the magnitude of the first excitation energy increases by large steps in the sequence (of N increasing) here given. On this last evidence one would be tempted to conclude that the odd-A xenon nuclei would be more polarizable than the corresponding barium nuclei. This is just the reverse of what is required.

It seems clear, then, that the problem of the isotope shift is a complicated one (and one factor which has not yet been taken into account theoretically for the odd-A species is that arising from the departure from spherical symmetry of charge, as expressed by the nuclear quadrupole moment), but it may be held that the results for $\binom{e}{e}$ species at least remain significant, albeit not wholly understood. If this is conceded then we must conclude that a discontinuity in structure is established between neutron numbers 88 and 90—though there is no sign of any corresponding discontinuity between $Z=88$ and $Z=90$ from our study of the energies and periods in α-disintegration (vide supra).

A method of investigation which promises well for the identification of minor discontinuities in relation to N has been developed by Hughes and collaborators (Hughes, Spatz and Goldstein, 1949; Hughes and Sherman, 1950). The effective neutron capture cross-section is measured for individual species by the activation method, using the secondary neutrons of fission for the investigation. These neutrons are characterized by an energy spectrum covering a range from zero to several MeV., with an effective energy of about 1 MeV. In such a measurement the experimental cross-section is determined largely by the density of energy states of the product nucleus over the range of excitation energies involved—and these energies

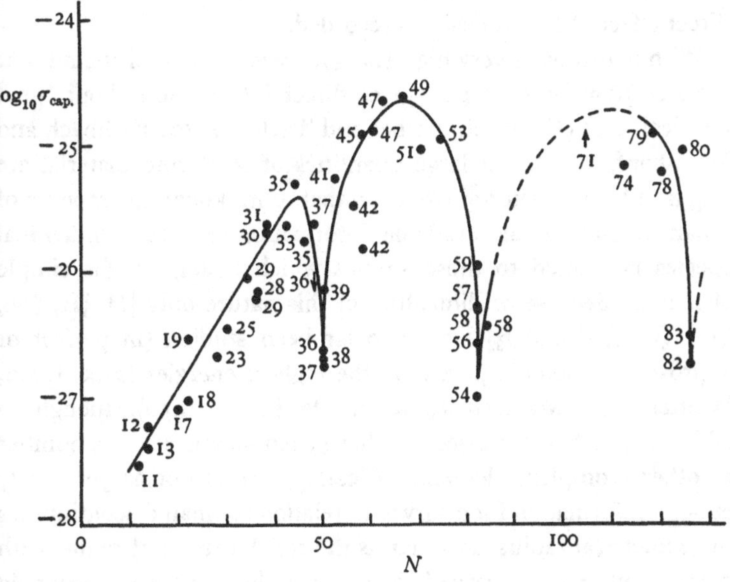

Fig. 14. Capture cross-section for fission neutrons as a function of N. $\sigma_{cap.}$ in cm.^{2}, Z values indicated on figure.

are greater or less as the energy of binding of the captured neutron is greater or less. Consequently small capture cross-sections are to be expected when the added neutron is the first (loosely-bound) neutron outside a closed nuclear shell. Results available up to date, and plotted on fig. 14, show how clearly this expectation is realized in relation to the capture cross-sections of species for which $N=50$, 82 and 126. There is also some suggestion of minor discontinuities at $N=56$, 70 and 112, the last two of which have previously been noted by other authors from a consideration of the variation of nuclear magnetic moment with N and Z (see Béné, Denis and Extermann, 1950).

Discontinuities have also been noted in plots of thermal neutron capture cross-sections against A, N and Z (first by Sinma and

Yamasaki, 1941), but these are less revealing than those just considered, precisely because the range of excitation energies of the product nucleus of thermal neutron capture is so much less than the corresponding range effective for the capture of fission neutrons. Usually a single excitation level is predominantly responsible for thermal neutron capture, whereas in the other case a much more direct effect of level-density is recorded.

With neutrons of very high energy, measurement of the nuclear cross-section for scattering gives direct information about 'geometrical' size (Fernbach, Serber and Taylor, 1949; Feshbach and Weisskopf, 1949), but large quantities of scattering material are required for the determination, so that, until kilogram amounts of separated isotopes are available, information relative to individual species is limited to those species which belong to the simple elements. Because of difficulties of this nature only ${}^{1}_{1}H$, ${}^{2}_{1}H$, ${}^{9}_{4}Be$, ${}^{12}_{6}C$, ${}^{16}_{8}O$, ${}^{27}_{13}Al$ and ${}^{238}_{92}U$ have so far been studied (in perfect or approximate isotopic purity) at the highest energies (Fox, Leith, Wouters and MacKenzie, 1950; De Juren, 1950), though in addition effective cross-sections have been measured for a number of other (complex) elements. Clearly there is not as yet nearly enough information for a survey in relation to small discontinuities in geometrical radius; in so far as the regular trend of radius with mass number is concerned, however, indications are reasonably definite. The most accurate measurements, at a neutron energy of 270 ± 30 MeV., are best fitted by the expression

$$\rho_s = 0{\cdot}127 A^{\frac{1}{3}} - 0{\cdot}082, \tag{2.16}$$

giving ρ_s, the geometrical radius, in units of 10^{-12} cm., in terms of A. Whilst there is no obvious reason for the exact identification of ρ_s of (2.16) with ρ_0 of (2.7), it is interesting to remark that, for $A = 209$, (2.16) gives $\rho_s = 0{\cdot}671$, which certainly falls within the limits of normality suggested by fig. 10 referring to ρ_0. Also, the trend of fig. 10, for $A > 230$ is best represented by $\rho_0 = 0{\cdot}137 A^{\frac{1}{3}}$, an expression in which the multiplying constant is not very different from that of (2.16). As indicating the degree of uncertainty involved in the interpretation of the cross-section measurements, it might be stated that earlier measurements using neutrons of about 90 MeV. energy (Cook, McMillan, Peterson and Sewell, 1949;

De Juren and Knable, 1950) were interpreted by the investigators in terms of an expression tending to $\rho_s = 0{\cdot}137A^{\frac{1}{3}} + 0{\cdot}050$, for large values of A.

2.6. Radioactive series branching and the systematics of 'pure α-active' species

The occurrence of $\alpha - \beta$ branching in radioactive series (Soddy, 1909; Fajans, 1912) is evidence for the fact that for certain nuclei the probabilities of the transformations

$$\binom{A}{Z} \xrightarrow{\alpha} \binom{A-4}{Z-2} \quad \text{and} \quad \binom{A}{Z} \xrightarrow{\beta} \binom{A}{Z+1}$$

are not of very different orders of magnitude. Obviously there will be other nuclei for which both transformations are energetically possible, but for which the disintegration probabilities are so widely different that, effectively, disintegration in a single mode is all that is observed. Generally speaking, the energy-dependence of disintegration probability is so much more rapid for α-disintegration than for β-disintegration that failure to observe β-disintegrations which are energetically possible must be attributed to the competition of α-processes of high energy,† whilst failure to observe possible α-disintegrations is due to the smallness of the energy of α-emission in the cases concerned. When we are dealing with species having a restricted range of Z values, we may say that $\alpha - \beta$ branching will be observed only if the energy of α-disintegration itself lies within reasonably narrow limits. For the heavy radioelements the upper and lower limits may conveniently be set at 7 and 5 MeV., respectively. On the other hand, there are many α-emitters amongst the classical radioelements which do not exhibit the phenomenon of branching, even though the energy of α-disintegration lies within these limits. We are led, therefore, to the concept of the 'pure α-active' species (Turner, 1940; Nordström, 1950; Wapstra, 1952), that is the unstable species which is stable against β- (or capture-) transformation and unstable only in relation to α-emission. We shall see that such a species is the counterpart, for values of Z greater than 83, of the 'truly' stable species of smaller Z

† We are neglecting, for the present, the dependence of disintegration constant on any factor other than disintegration energy. It is, of course, conceivable that an energetically possible β-disintegration is unobserved because of the magnitude of the spin change involved (see § 3.1).

—and we shall find that the systematics of such species are precisely the same as the systematics of the stable species which have been discussed at length in Chapter I.

In order to decide which are the pure α-active species we make use of fig. 12, extrapolating, where necessary, the curves of η against Z for different values of $A-2Z$. As an example of the results of this procedure, fig. 15 refers to the uranium series of elements.† The full lines of this figure represent the α- and β-disintegrations which have been observed and of which the energy release is known. The experimental values are given (in MeV.) in the figure. Vertical dotted lines represent other possible α-disintegrations, and the values of the energy shown opposite these lines have been obtained, by interpolation and extrapolation, from fig. 12, as already described. These more doubtful cases are indicated by the bracketing of energy values. Clearly, the information contained in fig. 15—if it be accurate enough—is sufficient to determine the stability or otherwise against β- and capture-transformation of the α-active species of the sequence ${}^{234}_{92}$U to ${}^{218}_{84}$Ra A, as also those of the recently observed sequence ${}^{230}_{92}$U to ${}^{214}_{84}$Ra C′ (Studier and Hyde, 1948), and of the hitherto unreported sequence ${}^{234}_{93}$Np to ${}^{218}_{85}$At. The basis upon which such determinations are possible is, of course, the principle of conservation: if there are alternative processes leading from a given initial to a given final nucleus, then the energy released in the two processes (or series of processes) must be the same. On this basis, for example (see Feather, 1945), we can admit the possibility of $\alpha-\beta$ branching with Ra A (it has been claimed (Karlik and Bernert, 1943*a*, 1944, 1946) that the α-particles of ${}^{218}_{85}$At have been observed having the predicted energy, and with small intensity, in sources of radium active deposit), but β-emission by ${}^{222}_{86}$Rn or any higher member of this series of α-bodies ($A-2Z=50$) is clearly impossible. As concerns capture-stability, the fact that ${}^{234}_{91}$U Z is β-active,‡ implies that ${}^{234}_{92}$U is capture-stable, and since η for ${}^{234}_{91}$U Z is almost certainly less than η for ${}^{234}_{92}$U, ${}^{230}_{89}$Ac is also β-active—and with greater energy release than is the case for ${}^{234}_{91}$U Z. But ${}^{230}_{89}$Ac being β-active, ${}^{230}_{90}$Io must be capture-stable. A like conclusion applies

† A similar figure has been given by Hall and Templeton (1950).

‡ It is now known that the ground state of the species $\binom{234}{91}$ is the isomer UZ (Feather and Bretscher, 1938; Bradt and Scherrer, 1945).

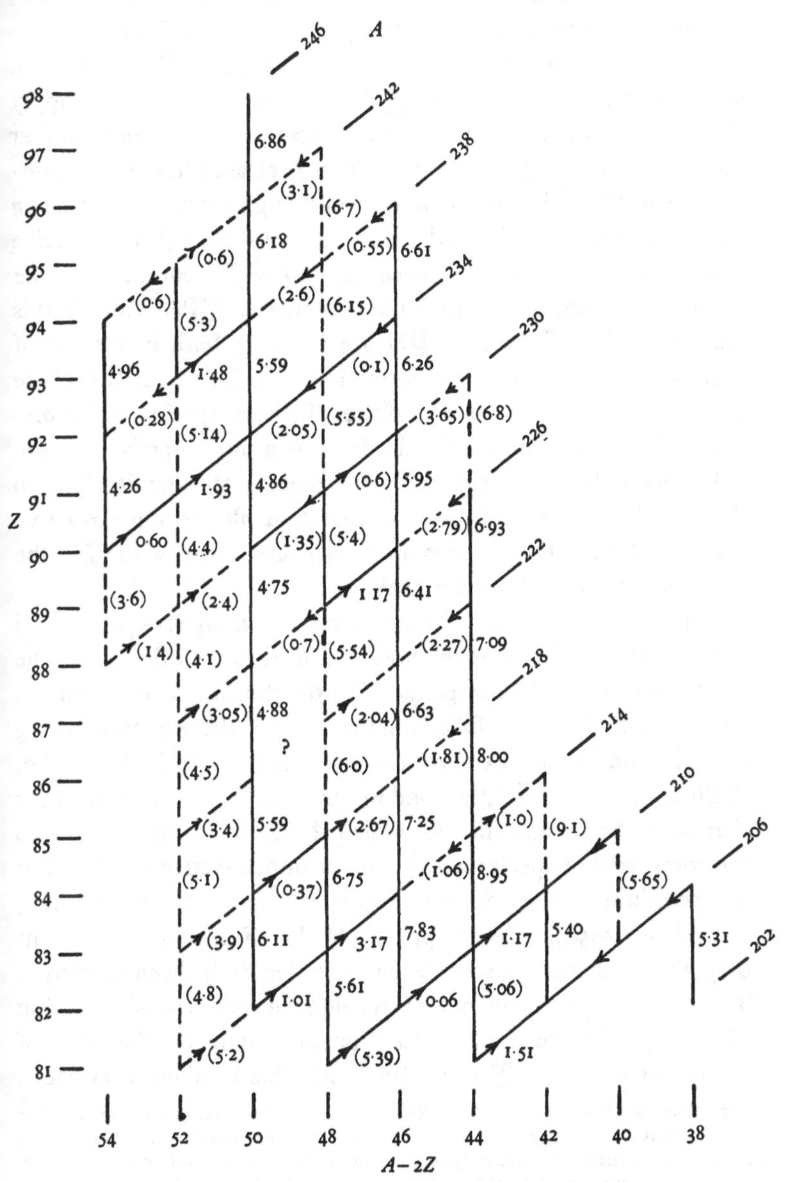

Fig. 15. Disintegration series, $A=4n+2$.

as far down the series as $^{218}_{84}$RaA: by this stage stability against capture transformation is by a margin of about 3·9 MeV.

We have shown, then, that the species $^{234}_{92}$U, $^{230}_{90}$Io, $^{226}_{88}$Ra and $^{222}_{86}$Rn are pure α-species—and that $^{218}_{84}$RaA is not. We can now apply similar methods to the sequence of species, with isotopic number 48, from $^{234}_{93}$Np to $^{218}_{85}$At. Since the energy released in the β-disintegration of $^{214}_{83}$RaC is considerable, even though for pairs of isobars η is greater when $A-2Z=46$ than it is when $A-2Z=48$, the α-active species of the latter sequence ($A-2Z=48$) continue to be β-unstable at least as far up in the sequence as $^{230}_{91}$Pa. Thus, in this sequence, $^{230}_{91}$Pa, $^{226}_{89}$Ac and $^{222}_{87}$Fr are unstable both in respect of β-emission and capture-transformation, and $^{218}_{85}$At and $^{214}_{83}$RaC are β-active and capture-stable. None of these species, therefore, having isotopic number 48 and odd Z, is a pure α-active species.

In much the same way, for the sequence of α-bodies $^{230}_{92}$U to $^{214}_{84}$RaC′, having $A-2Z=46$, we find that all are pure α-active species. β-stability increases as A increases, until with $^{230}_{92}$U the stability margin is about 3·7 MeV.

In fig. 15 we have taken the uranium series as an example worked out in detail. If the same method is applied to the other series the whole picture may be completed. For the thorium series (Feather, 1948*a*) we find that the following members of the sequence having isotopic number 48 are pure α-bodies, namely $^{216}_{84}$ThA,† $^{220}_{86}$Tn, $^{224}_{88}$ThX, $^{228}_{90}$RdTh and $^{232}_{92}$U—and in consequence we must conclude that the isobaric sequences for which $A-2Z=50$ and $A-2Z=46$ are comprised of species which are not pure α-active species. For the actinium series (see Schintlmeister, 1938; Vigneron, 1947; Jentschke, 1950), $^{215}_{84}$AcA proves to be β-unstable by about 0·75 MeV., though β-activity has not definitely been observed. This is not surprising in view of the high energy of α-disintegration of this species (maximum β:α branching ratio on the basis of fig. 17 $\sim 6\times10^{-7}$).‡ $^{219}_{86}$An is also β-unstable by about 0·25 MeV.

† This conclusion throws doubt (Flügge and Krebs, 1944) on the attribution (Karlik and Bernert, 1943*b*, 1944) of the reported group of α-particles of long range from sources of ThA to $^{216}_{85}$At formed in the β-disintegration of this body.

‡ An observation of Karlik and Bernert (1944) interpreted by them as due to $\beta-\alpha$ branching of AcA, with branching ratio 5×10^{-6}, can be fully accepted only if it is assumed that AcA is β-unstable to the extent of at least 1·4 MeV. A more recent report by Avignon (1950), placing the branching ratio at $(2{\cdot}3\pm0{\cdot}2)\times10^{-6}$, reduces the extent of the discrepancy.

on the evidence available, though, again, β-activity has not been detected. ${}^{223}_{88}$AcX and ${}^{227}_{90}$RdAc, on the other hand, are clearly β-stable. Since these latter species are themselves daughter products in β-disintegration (AcX being produced in the β-disintegration of ${}^{223}_{87}$AcK) they are also capture-stable, and are thus pure α-species. The isotopic number of the four α-bodies which we have just been discussing is 47; concerning the isobaric series from ${}^{227}_{89}$Ac to ${}^{215}_{83}$Bi, of isotopic number 49, it turns out that all are β-unstable. But ${}^{231}_{91}$Pa ($A-2Z=49$) is a pure α-species, and the next higher member, ${}^{235}_{93}$Np, being capture-active (Magnusson, Thompson and Seaborg, 1950), is not. By the same token, however, ${}^{235}_{92}$U ($A-2Z=51$) is a pure α-body. In the sequence for which $A-2Z=45$, ${}^{227}_{91}$Pa and ${}^{223}_{89}$Ac are capture-active (Ghiorso, Meinke and Seaborg, 1948), ${}^{219}_{87}$Fr and ${}^{215}_{85}$At are probably both β- and capture-stable and ${}^{211}_{83}$AcC is β-active. On the other hand, ${}^{211}_{84}$AcC′ ($A-2Z=43$) is certainly a pure α-body, being formed in the β-disintegration of AcC and also in the capture-transformation of ${}^{211}_{85}$At. Finally, in the neptunium ($A=4n+1$) series, it would appear that the pure α-species are ${}^{237}_{93}$Np with $A-2Z=51$, ${}^{233}_{92}$U and ${}^{229}_{90}$Th, with $A-2Z=49$, ${}^{225}_{89}$Ac with $A-2Z=47$, and ${}^{221}_{88}$Ra, ${}^{217}_{86}$Em and ${}^{213}_{84}$Po with $A-2Z=45$.

We may summarize these conclusions and others, some of them still more or less provisional, in Table XI (cf. Perlman, Ghiorso and Seaborg, 1950). Cases of doubt are here indicated by bracketed A-values. In fig. 16 the systematics of stable and pure α-active species (taken from Table XI) are represented in one diagram, of the same type as fig. 1 ($A/(A-2Z)$), over the range of mass numbers from 190 to 244 (cf. Nordström, 1950). Full circles are employed for ‘certain’ and open circles for ‘doubtful’ species. The general aspect of this figure substantiates our claim (p. 83) that for $Z>83$ the pure α-active species is the exact counterpart of the stable species of smaller Z, having precisely the same existence rules. Thus, in fig. 16, pure α-species of odd Z have odd A also, and neither for any odd nor for any even Z are there more than two pure α-species of odd A. When there are two such species (e.g. for $Z=90$), as before they have neighbouring odd values of A.

From the general aspect of fig. 16 we can draw certain conclusions regarding the doubtful cases listed in Table XI. It would thus appear that the species $^{212}_{84}$Po, $^{216}_{86}$Em, $^{220}_{88}$Ra and probably $^{224}_{90}$Th ($A-2Z=44$) are pure α-species, and that the same is true of $^{208}_{84}$Po ($A-2Z=40$) and $^{214}_{86}$Em ($A-2Z=42$). Similarly, $^{232}_{90}$Th and $^{238}_{92}$U can hardly fail to belong to this class, nor the sequence $^{236}_{92}$U, $^{240}_{94}$Pu and $^{244}_{96}$Cm, which species, along with $^{232}_{90}$Th, have $A-2Z=52$. Some doubt remains, perhaps, regarding $^{242}_{94}$Pu, but $^{242}_{96}$Cm is probably both β- and capture-stable.

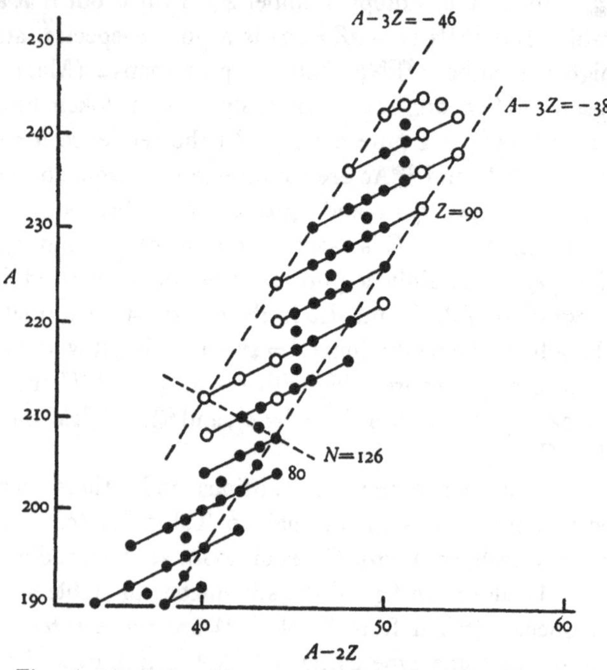

Fig. 16. 'Stable' and 'pure α-active' species, $190 \leqslant A \leqslant 244$.
● certain, ○ doubtful.

If we accept the above conclusions, further interesting results may be derived by an argument which is the converse of that used in deducing them from the facts as they are known. Here we shall confine ourselves to a single example, eliciting the results which follow from the assumption that $^{208}_{82}$Pb, $^{208}_{84}$Po, $^{212}_{84}$Po and $^{212}_{86}$Em are all 'stable' or pure α-active species. If this assumption is accepted then both $^{208}_{83}$Bi and $^{212}_{85}$At are both capture- and β-unstable. Neither species has as yet been definitely identified, and as regards $^{208}_{83}$Bi the experimental result is that it has a half-value period either less than

TABLE XI. *'Stable' and pure α-active species*

Z	A
82	204, 206, 207, 208
83	209
84	(208), 210, 211, (212), 213, 214, 216
85	215
86	(212), (214), (216), 217, 218, 220, (222)
87	219
88	(220), 221, 222, 223, 224, 226
89	225
90	(224), 226, 227, 228, 229, 230, (232)
91	231
92	230, 232, 233, 234, 235, (236), (238)
93	237
94	(236), 238, 239, (240), (242)
95	241, (243)
96	(242), (243), (244)

30 sec. or greater than 100 years (Neumann, Howland and Perlman, 1950). So far as its α-disintegration properties are concerned, a lifetime of 100 years or more would appear reasonable, but we have to consider its other instabilities as well—particularly in relation to the α-disintegration energies of $^{212}_{85}$At and of $^{208}_{83}$Bi itself. The α-distintegration energies of $^{212}_{84}$Po and $^{212}_{86}$Em are 8·95 and 6·30 MeV., respectively. Clearly, the larger the α-disintegration energy of $^{212}_{85}$At (or the smaller that of $^{208}_{83}$Bi) the smaller are the margins of instability of $^{208}_{83}$Bi in respect both of capture-transformation and of β-disintegration. Again, if the lifetime of $^{208}_{83}$Bi were in fact less than 30 sec. this would almost certainly be because the β-disintegration energy was of the order of 3 MeV. (and the spin change was favourable, see fig. 17) rather than because the energy for K-capture was high. But in either case it would require the α-disintegration energy of $^{212}_{85}$At to be of the order of 6 MeV. or less and the first assumption would require that of $^{208}_{83}$Bi to be of the order of 7 MeV. or more. The former seems unlikely in view of the regularities of fig. 12 and the latter is altogether impossible; it would appear much more probable that the α-disintegration energy of $^{212}_{85}$At is greater than 7 MeV., the β-integration energy of $^{208}_{83}$Bi is small (it might even be negative, see Templeton, 1950) and the K-capture transformation of this species sufficiently forbidden to be of long lifetime. Disintegration energies of 5·4 MeV. (experimental) 7·3 and 9·1 MeV. (both estimated) for α-disintegration for the three species $^{210}_{84}$Po, $^{212}_{85}$At and $^{214}_{86}$Em would

satisfactorily define the rising portion of the curve for $A-2Z=42$ in fig. 12, and the point for the last mentioned species (having $N=128$) would be expected to fix the maximum of this curve. On this basis $^{214}_{85}$At would be β-unstable to the extent of 1·0 MeV., but the $\beta:\alpha$ branching ratio would still be too low for the fact of branching to be experimentally confirmed.

Finally, we may compare fig. 16 more generally with fig. 1. We find, first of all, that the discontinuity at $N=126$ does not show up anything like so clearly in the former figure as do the discontinuities at $N=50$ and $N=82$ in the latter. This appears to be connected with a 'narrowing of the valley on the energy surface' towards the highest 'regions' of A and Z (Wapstra, 1950); for no value of N are there more than four points on fig. 16, compared with seven for $N=82$ and five for $N=50$ on fig. 1. The lower stability limit in A, for Z constant (or the higher limit in Z, for N (or A) constant) appears to change by two units above $N=126$, but there is no extension of stability towards lower Z for this value of N. $^{206}_{80}$Hg is unknown. As regards unfavoured N values, $N=128$ shares with $N=120$ the distinction of being represented by two points, only, on fig. 16: $^{210}_{82}$Ra D is just β-unstable, and $^{202}_{82}$Pb is unknown (Duckworth, Black and Woodcock, 1949). In both cases features of instability connected with neutron binding obviously outweigh the tendency towards stability due to the closure at $Z=82$ of a proton shell. For $^{210}_{82}$Ra D this is perhaps not surprising; that it should be the case for $^{202}_{82}$Pb is less understandable and appears to indicate that a subsidiary discontinuity in neutron binding occurs at $N=118$. No such discontinuity could be predicted on the basis of fig. 1; between $N=100$ and $N=126$ the only suggestion of neutron shell closure occurs at $N=106$. On fig. 16 evidence for possible subsidiary discontinuities beyond $N=126$ is slight. The most likely would appear to be at $N=136$. The fact that no discontinuities are evident in the systematics of points representing the species of odd A on fig. 16 (not even the discontinuity at $N=126$) adds strength to the conclusion that no significant reorganization of nuclear structure occurs in the range of the classical radioelements and the earlier transuranics. We have previously described the adjustments beyond $N=50$ and $N=82$ as a loosening of structure whereby additional positive charge could be added without imperilling stability. It would seem that such an adjustment is no longer possible to the same extent when the total charge has increased to 80 or 90 units. There is nothing intrinsically improbable in this conclusion: ultimately (as Z increases) spontaneous fission must be taken into account (Turner, 1945) as a fourth type of transformation which has a definite bearing upon the lifetime of any nuclear species.† When spontaneous fission becomes 'instantaneous', even for the ground state of such a species—as it is for the excited state of the compound nucleus in neutron-induced fission—a limit to the process of nuclear synthesis will clearly have been reached. At that stage the number of the elements (Feather, 1946) will have taken on the aspect of a known integer—an ascertained constant of nature (see Chapter IV).

† Hanna, Harvey, Moss and Tunnicliffe (1951) have shown that the instability of $^{242}_{96}Cm$ towards spontaneous fission is such that this species would be 'missing' on earth to-day even if it were stable except for spontaneous fission (effective lifetime 7×10^6 years). Actually it is α-active of half-value period 162·5 days.

CHAPTER III

REGULARITIES IN β-DISINTEGRATION

3.1. The Sargent diagram

We have already noted that the variation of disintegration constant with disintegration energy is much less rapid in β-disintegration than it is in α-disintegration (p. 83). It is also a fact that the dependence of disintegration probability on the magnitude of the change in nuclear spin is more marked in β-disintegration than in α-disintegration. As a consequence of these two facts excitation of the residual nucleus is considerably more common with the former process than it is with the latter—that is, β-active bodies emit γ-rays, particularly γ-rays of high energy, much more frequently than α-active bodies do. Not infrequently, in β-disintegration the residual nucleus is left excited with effectively unit probability, an extremely rare occurrence in α-disintegration: in such cases the effect of a favourable spin-change must be considered to have outweighed entirely the effect of an unfavourable balance of energy in determining the relative transition probabilities for ground-to-excited state and ground-to-ground state transitions. When this occurs one result of some practical importance is that the quantum energy of the γ-rays, as well as the kinetic energy of the β-particles, must be determined before the total disintegration energy of the species is known by direct experiment.

Experimentally, the survey of β-disintegrations is also rendered difficult by the now well-known feature that even in the simplest case (when one initial and one final nuclear state, only, are involved) the energy spectrum of the disintegration electrons is a continuous spectrum. In the more general case, therefore, the problem of the elucidation of the fine structure in α-particle spectra is paralleled by the much more difficult problem of the analysis (Ellis and Mott, 1933) of the 'complex' continuous spectrum of β-particle energies into 'partial' continuous spectra. Clearly, no empirical regularity between disintegration constant and disintegration energy has any real basis of significance until a considerable body of results so analysed is available for the systematizer.

But progress in science does not always wait on the preparation of secure bases of real logical significance, and it is a fact of history that the Sargent diagram (Sargent, 1933) was put forward as exhibiting an obviously significant empirical regularity long before any analyses of complex into partial continuous spectra had been carried out. It was put forward at a time when there was no generally accepted theory of the disintegration process, and when even the view that the end-point energy of the continuous β-spectrum (E_0) is the unique transition-energy characterizing a well-defined disintegration mode was still in dispute. Its success helped to establish the principle of the importance of the end-point energy, but from the vantage point of to-day we are able to appreciate much more clearly than was possible eighteen years ago the combination of circumstances which made that success possible. In 1933 the only β-active bodies known were those of 'classical' radioactivity, effectively a compact group in relation to atomic number Z, and these included a sufficient number of examples of bodies having nearly simple energy spectra (bodies transforming predominantly by a single mode in each case) for a significant regularity to emerge. The original diagram included only 12 plotted points. One of these (for $^{211}_{82}$Ac B) appeared obviously anomalous—and was before long (Sargent, 1939) shown to be incorrectly placed—and the other 11 were held to exhibit a distinction between 'allowed' and 'forbidden' disintegration transitions which cut across the natural affiliation of the β-active bodies in the three series. This was a bold generalization on meagre evidence, but in the upshot it has been largely vindicated: subsequent developments have altered matters of detail rather than fundamentals, and the allowed/forbidden classification has remained a significant feature of all later theorizing. With the taking in of more extensive experimental data, and the realization that the empirical classification in respect of degree of forbiddenness is broadly a classification in respect of change of nuclear spin (Fermi, 1934), the way was clear for the recognition of other lines on the diagram ('second-forbidden', 'third-forbidden' lines, etc.), but it is a valid point to make that the fate of the original two Sargent lines has been entirely different from that of the three lines of the first Geiger-Nuttall diagram. As we have already seen, the original Geiger-Nuttall lines became obliterated as insignificant in the subsequent treatment of the data; the Sargent lines have been retained, and have been added to by a natural extension of the interpretative scheme as more information has become available as the result of experiment.

The ready availability of 'additional' information of a particular type was first clearly demonstrated by Gamow and Teller (1936). These authors drew attention to the fact that the 'unobserved' transitions, characterizing the energetically well-allowed ground-to-ground disintegration modes of those bodies which give rise to daughter products showing sensibly 100% nuclear excitation, are merely transitions of a high degree of forbiddenness. Determination of the energies available for β-disintegration in these modes (see above), and of the upper limits of the corresponding disintegration probabilities, provides information from which 'limiting' points on a Sargent diagram may be plotted. Such points will obviously give some indication of the positions of the multi-forbidden lines on the diagram.

Figs. 17 and 18 are Sargent diagrams, for the heavy β-active bodies having $80 \leqslant Z \leqslant 84$, and $88 \leqslant Z \leqslant 93$, containing some of these 'limiting' points and generally constructed after the pattern of the diagram published by Feather and Bretscher in 1938. The point of this construction is the method of representation of the degree of uncertainty in the data employed. When weak sources and the absorption method have been used in deducing β-particle energies there may have been a systematic under-estimation of energy, similarly a systematic over-estimation of disintegration probability is always possible when a complex spectrum has not been fully analysed, or has been assumed to be simple. These possibilities are represented by unsymmetrical 'loops of uncertainty' on the diagram. For the 'unobserved' transitions, obviously, the loops of uncertainty are of indefinite extent in the direction of $\log \lambda$ decreasing. Symmetrical loops of uncertainty represent experimentally determined probable errors in the usual way.

Although there is now considerably more information concerning artificially produced β-emitters than there is concerning the β-emitters of classical radioactivity (and figs. 17 and 18 include many points for artificially produced β-emitters in the ranges of atomic number to which they apply), there is good reason, over and

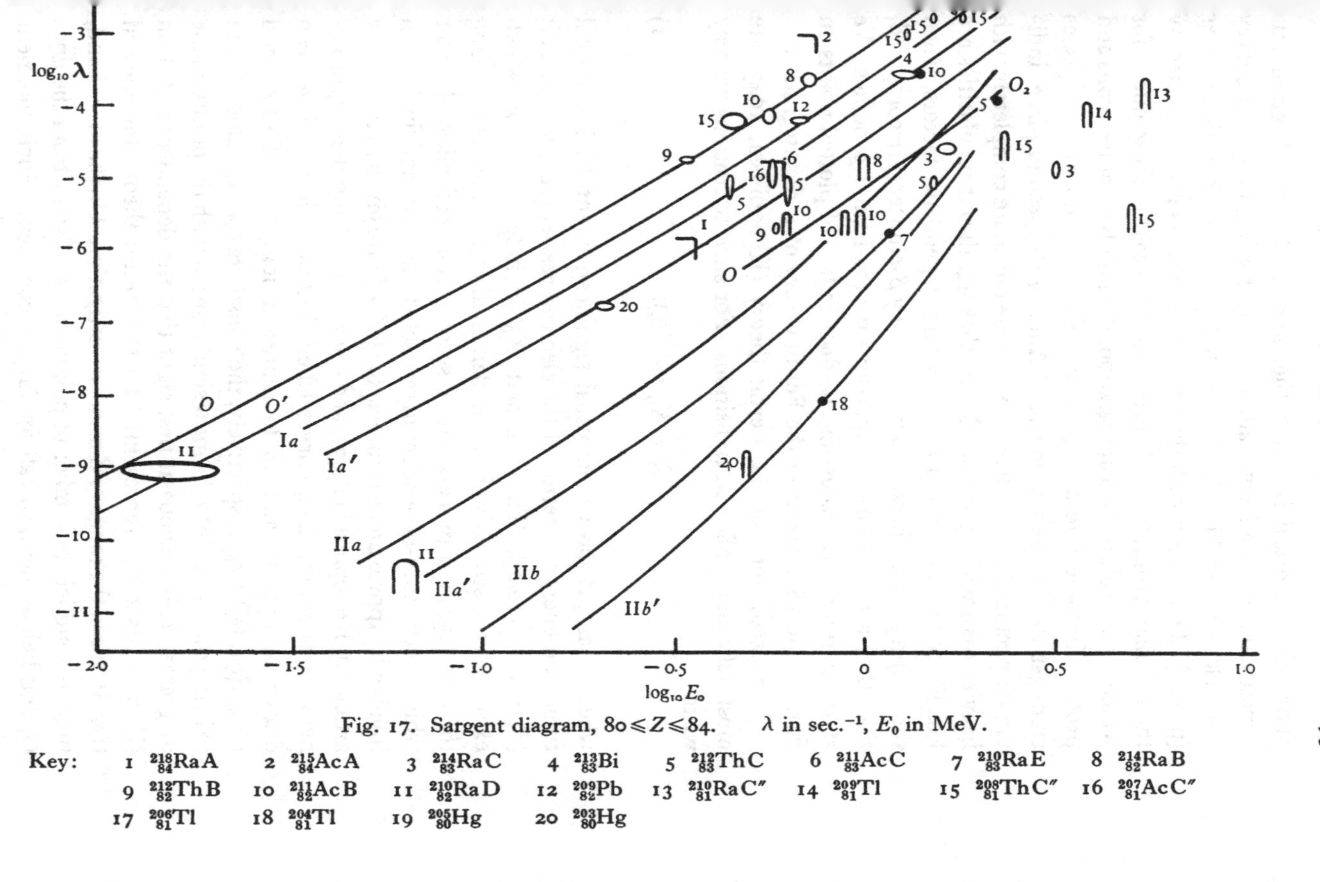

Fig. 17. Sargent diagram, $80 \leqslant Z \leqslant 84$. λ in sec.$^{-1}$, E_0 in MeV.

Key: 1 $^{218}_{84}$RaA 2 $^{215}_{84}$AcA 3 $^{214}_{83}$RaC 4 $^{213}_{83}$Bi 5 $^{212}_{83}$ThC 6 $^{211}_{83}$AcC 7 $^{210}_{83}$RaE 8 $^{214}_{82}$RaB 9 $^{212}_{82}$ThB 10 $^{211}_{82}$AcB 11 $^{210}_{82}$RaD 12 $^{209}_{82}$Pb 13 $^{210}_{81}$RaC″ 14 $^{209}_{81}$Tl 15 $^{208}_{81}$ThC″ 16 $^{207}_{81}$AcC″ 17 $^{206}_{81}$Tl 18 $^{204}_{81}$Tl 19 $^{205}_{80}$Hg 20 $^{203}_{80}$Hg above the historical fact that the original Sargent diagram was constructed for the heavy radioelements, for beginning our survey with elements of high Z. In the first place, as we shall see later (p. 106), the position with the lightest β-emitters is complicated by the peculiar properties of 'mirror' nuclei—that is those nuclei for which, as a result of β-disintegration, the numbers of neutrons and protons become interchanged—and there are no mirror species amongst the heaviest β-emitters. And, in the second place, individual peculiarities are likely to average out more completely with heavier than with lighter nuclei, so that significant regularities are less likely to be obscured by 'structure-dependent' factors, when the disintegration characteristics of heavy β-emitters are compared, than they are in similar comparisons for the lighter β-bodies. We start, therefore, by discussing the distribution of plotted points on figs. 17 and 18 in terms of the Fermi theory of β-disintegration.

The predictions of the Fermi theory (Konopinski, 1943) are most definite for allowed transitions. For such transitions we may write

$$\lambda_0 = K_0 f(Z, W_0), \qquad (3.1)$$

where λ and Z have their usual significance, and W_0 is the total transition energy in terms of the electron rest-energy, mc^2, as unit $(E_0 = (W_0 - 1)mc^2)$. The constant K_0 includes a factor which expresses the strength of the fundamental interaction determining the disintegration process, and a second more complicated factor which is structure-dependent in respect of the initial and final nuclei involved. Approximate formulae for $f(Z, W_0)$ are given in the review article by Konopinski (1943), and elsewhere. Using these approximations as appropriate, curves labelled O in each case have been drawn in figs. 17 and 18 to represent $\log_{10} f(82, W_0) + c_{82}$ and $\log_{10} f(91, W_0) + c_{91}$ respectively, the constants c_{82}, c_{91} being chosen to bring the curves into the most significant fit with the points on the diagrams. It is obvious at first sight that the dependence of λ on W_0 is satisfactorily represented by the Fermi theory for allowed transitions in the range of Z under investigation. On the other hand, there is a sufficient number of points not far distant from the lines O, and below them, to call for further comment. Such comment discloses a result which applies throughout the whole of our survey.

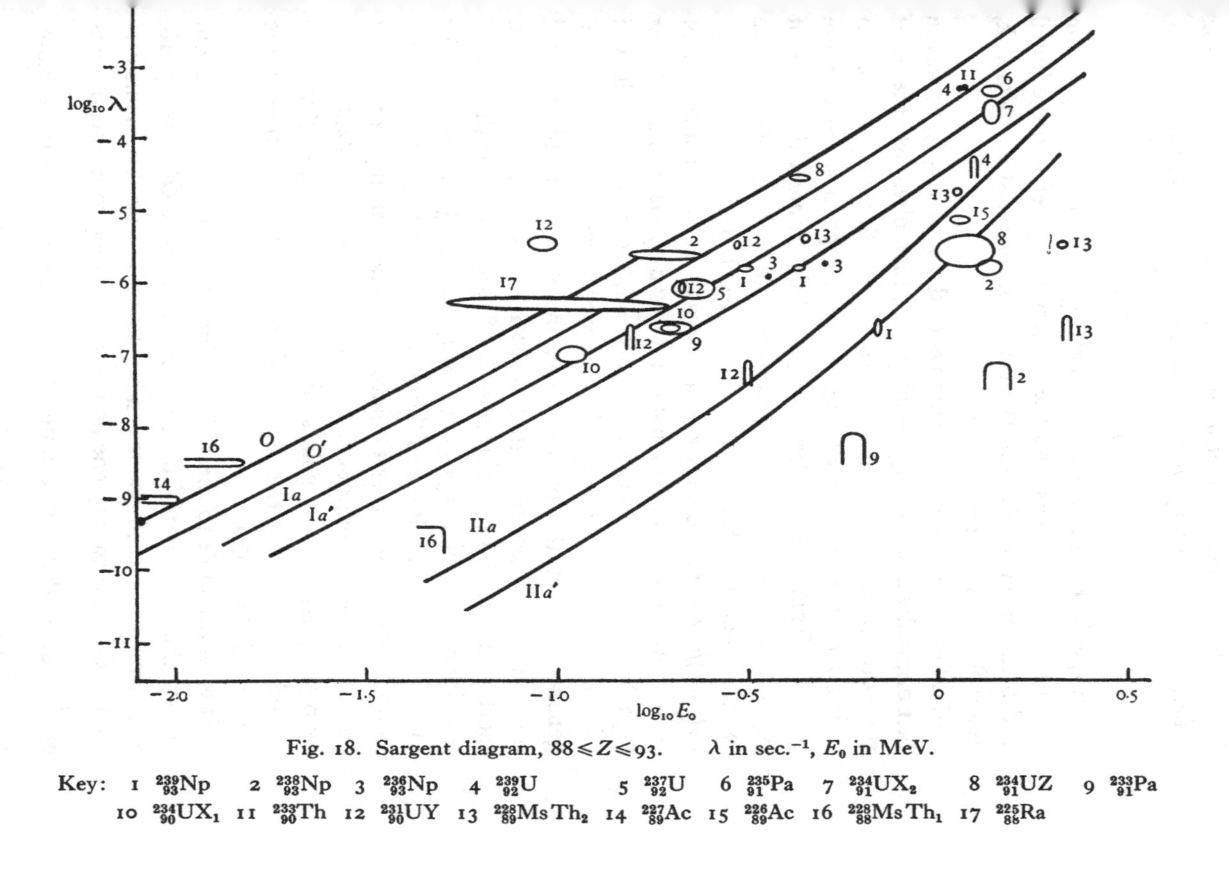

Fig. 18. Sargent diagram, $88 \leqslant Z \leqslant 93$. λ in sec.$^{-1}$, E_0 in MeV.

Key: 1 $^{239}_{93}$Np 2 $^{238}_{93}$Np 3 $^{236}_{93}$Np 4 $^{239}_{92}$U 5 $^{237}_{92}$U 6 $^{235}_{91}$Pa 7 $^{234}_{91}$UX$_2$ 8 $^{234}_{91}$UZ 9 $^{233}_{91}$Pa
10 $^{234}_{90}$UX$_1$ 11 $^{233}_{90}$Th 12 $^{231}_{90}$UY 13 $^{228}_{89}$MsTh$_2$ 14 $^{227}_{89}$Ac 15 $^{226}_{89}$Ac 16 $^{228}_{88}$MsTh$_1$ 17 $^{225}_{88}$Ra

This is the general statistical result that the constant K_0 contains a factor which is unity when $I_i \geqslant I_f$ but assumes the value

$$(2I_f+1)/(2I_i+1)$$

when $I_i < I_f$. Here I denotes the total angular momentum quantum number of the nucleus, and the subscripts i, f refer to the initial and final nuclei involved in the transition in question (Marshak, 1942; Strachan, 1948).

If allowed transitions were always of the type $I_i = I_f$, as was in fact originally assumed by Fermi (1934), then obviously the lie of the corresponding Sargent line would be effectively unique in a diagram covering a small range of Z, except in so far as more subtle differences in nuclear 'structure' determined the value of K_0 in each case. Such differences may indeed operate to some extent, even with highly complex nuclei, but the weight factor $(2I_f+1)/(2I_i+1)$ provides a more obvious cause for a scatter of points. According to views first put forward by Gamow and Teller (1936) certain transitions for which $I_i = I_f \pm 1$ are equally allowed with others having $I_i = I_f$. If these views are accepted, as they generally are at the present time, the most probable allowed transitions are those belonging to the $I_i \rightarrow I_f$ class $0 \rightarrow 1$, the next, most probable those of the class $\frac{1}{2} \rightarrow \frac{3}{2}$, and so on, with allowed transitions for which $I_i = I_f$ or $I_i = I_f + 1$ only $\frac{1}{3}$ as probable, for a given W_0, Z, as transitions of the class $0 \rightarrow 1$. Having regard to these results, subsidiary curves O' have been drawn in figs. 17 and 18 parallel to the theoretical curve O in each case and distant from it by 0·477 ($\log_{10} 3$). Any points falling on or between OO' we now consider as representative of allowed transitions on the basis of the Gamow-Teller rules.

It is worth while to check the position further at this stage. On fig. 17, thirteen points fall on or within the lines OO'; on fig. 18, eight similarly placed points can be identified with reasonable certainty. Of these twenty-one points, four belong to disintegrations of $\begin{pmatrix} e \\ e \end{pmatrix}$ nuclei, that is to disintegrations for which almost certainly $I_i = 0$. The most probable disintegration modes of each of the species $^{228}_{88}$Ms Th$_1$, $^{214}_{82}$Ra B, $^{212}_{82}$Th B and $^{210}_{82}$Ra D are here in question. Two of the four representative points of this group clearly lie on line

O rather than below it—and the same may well be true of the other two points, those for the 'observed' modes of Ra D and Ms Th_1. This is entirely as it should be; representative points for allowed transitions from $I_i = 0$ can lie only on O according to Gamow-Teller rules, for according to these rules transitions of the class $0 \rightarrow 0$ are not allowed. Other points lying almost certainly on O on figs. 17 and 18 may be grouped as follows: first, points representing transitions to high excited states of the nuclei $^{210}_{82}$Ra D, $^{209}_{82}$Pb and $^{208}_{82}$Pb; secondly, points representing transitions presumably to the ground states of the nuclei $^{234}_{92}$U, $^{227}_{90}$Rd Ac, $^{207}_{82}$Pb and $^{206}_{82}$Pb, respectively. In order to complete the picture we have to consider whether these transitions can possibly belong to the class for which $I_i < I_f$. Concerning transitions of the first group nothing definite can be said, but difficulties in respect of each of the four transitions of the second group must be noted here. For the nucleus $^{207}_{82}$Pb, $I = \frac{1}{2}$, and we should expect $I = 0$ for the nuclei $^{234}_{92}$U and $^{206}_{82}$Pb in the ground state. As regards $^{227}_{90}$Rd Ac we have predicted $I = \frac{11}{2}$ or $\frac{13}{2}$ on p. 60. If these values are accepted, and if the β-distintegrations in question (those of $^{207}_{81}$Ac C″, $^{234}_{91}$U X_2, $^{206}_{81}$Tl and $^{227}_{89}$Ac) are in fact ground-to-ground state disintegrations, obviously $I_i < I_f$ cannot be satisfied. It is not clear how the difficulty can be resolved without postulating the excitation of very low-lying (possibly metastable) states in the disintegrations concerned (for the case of UX_2 see Feather and Richardson (1948)).

Taking the points lying almost certainly on O' on figs. 17 and 18, something definite can be said regarding the disintegrations $^{205}_{80}$Hg $\rightarrow$ $^{205}_{81}$Tl, $^{209}_{82}$Pb $\rightarrow$ $^{209}_{83}$Bi and $^{208}_{81}$Th C″ $\rightarrow$ $^{208}_{82}$Pb* (two or three modes).† For the first of these $I_f = \frac{1}{2}$, so that $I_i \geqslant I_f$ as is required; for the second $I_f = \frac{9}{2}$, and it is plausible to assume that $I_i = I_f$; for the last group $I_i \geqslant 4$ (since the ground-to-ground state disintegration $^{208}_{81}$Th C″ $\rightarrow$ $^{208}_{82}$Pb is so highly forbidden), and $I_i \geqslant I_f$ is again a plausible assumption (see Petch and Johns, 1950). No obvious difficulty, then, arises here.

Having discussed the empirically allowed transitions,‡ we now turn to the more difficult consideration of those transitions which

† An asterisk, as here used, denotes an excited state of the nucleus concerned.

‡ The anomalously placed point for one of the transitions $^{231}_{90}$UY $\rightarrow$ $^{231}_{91}$Pa on fig. 18 is only recently reported (Freedman, Wagner, Jaffey and May, 1951, 1952) and possibly subject to doubt.

are forbidden—to the first or higher degree. The formal predictions of theory here are briefly as follows (Konopinski and Uhlenbeck, 1941). We note first that we can, with some error, but not too great a distortion of truth, write, for a transition which is nth forbidden,

$$\lambda_n = K_n f(Z, W_0), \tag{3.2}$$

the function $f(Z, W_0)$ being taken over without change from (3.1), and the constant K_n having the same general character as K_0, but being of a smaller magnitude.† In particular K_n contains the weight factor of Marshak, as does K_0. The nuclear radius being written as $\rho(h/2\pi mc)$, and α being the fine-structure constant $2\pi e^2/hc(1/137)$, we then have, except in certain special cases, when

$$\alpha Z \gg 2\rho W_0, \tag{3.3}$$

$$K_0/K_n \sim (\alpha Z/2)^{-2}(\rho W_0)^{-2(n-1)}, \tag{3.4}$$

or

$$K_0/K_n \sim (\rho W_0)^{-2n}, \tag{3.5}$$

depending on the change of spin in the transition. Since (3.3) is reasonably well satisfied for the majority of the transitions represented on figs. 17 and 18 (the mean value of $\alpha Z/2\rho$ for the species concerned being 13·7, and $W_0 = 13{\cdot}7$ corresponding to $E_0 = 6{\cdot}5$ MeV.), we have to face the general situation that the Sargent line (or 'band', compare the band OO' for allowed transitions) for nth forbidden transitions ($n > 0$) is unlikely to be uniquely placed on these figures. If the original selection rules of Fermi were to apply, only (3.4) would be operative, and the Sargent bands for the different values of n would each be unique, but with the Gamow-Teller rules both (3.4) and (3.5) are relevant and the Sargent bands are duplicated. In such a situation further interpretation of figs. 17 and 18 must obviously be tentative.

We shall consider first possible first-forbidden transitions for which (3.4) is valid. For $Z = 82$, $(\alpha Z/2)^{-2} = 11{\cdot}2$; for $Z = 91$ the corresponding value is 9·1. According to (3.4), therefore, a first-forbidden band parallel to OO' (fig. 17 or fig. 18) and distant from it about 1·0 in $\log_{10} \lambda$ is to be expected.‡ A survey of the figures

† In general K_n is $(1/ft)\log 2$, ft being the 'comparative half-life' introduced by Konopinski (1943) as a semi-empirical quantity in terms of which degrees of forbiddenness might be assigned without recourse to the Sargent diagram (see Feenberg and Trigg, 1950; Moszkowski, 1951; Feingold, 1951).

‡ Such a feature was originally postulated, on the basis of a purely empirical review of the data, by the writer (Feather, 1948 *b*) and was later (Feather and Richardson, 1948) related to theory in the manner given above.

shows that there are certainly several well-authenticated points not very far below the allowed band on each—and over the greater portion of the length of the band, taking the evidence of the two figures together. Now (3.4) with $n=1$ provides the only possibility of a ratio K_0/K_n which is independent of W_0. Fact would appear, then, to bear out prediction in this matter, and, as best satisfying the requirements of the experimental data, a first-forbidden band $I_a I'_a$ has been drawn in fig. 17, 0·85 in $\log_{10}\lambda$ below OO', and in fig. 18, 0·93 below. Since the permitted changes of spin for the transitions here represented are 0, ± 1, the width of the new first-forbidden bands is the same as that of the allowed bands OO'. In total, some fifteen points may plausibly be regarded as falling within the bands $I_a I'_a$ on the two figures together.

Continuing our survey, we omit at this stage any attempt to locate the Sargent lines I_b, I'_b which would refer to first-forbidden transitions for which (3.5) is valid, since no transition is known amongst the radioelements under discussion to be characterized by the appropriate values of change of spin and parity ($\Delta I = \pm 2$; parity change, 'yes'); we turn rather to the consideration of certain 'paired' disintegrations of a particular type. Such paired disintegrations are the successive disintegrations, ground-to-ground state in each case, of parent and daughter species when the initial parent is an $\binom{e}{e}$ species. The disintegration scheme

$$\binom{e}{e} \xrightarrow{\beta} \binom{o}{o} \xrightarrow{\beta} \binom{e}{e} \tag{3.6}$$

represents a successive pair of this type. The initial and final nuclei in such an event almost certainly have $I=0$ and even parity. The two disintegrations, therefore, are characterized by equal but opposite values of ΔI and the same parity change: they should be forbidden to identical degree. More precisely, since the Marshak weight factor would be operative for one disintegration and not for the other, the two representative points of the pair should lie on the same Sargent band, but on opposite sides of the band. Amongst the disintegrations now under discussion the pairs are ${}^{234}_{90}\mathrm{UX}_1 \rightarrow {}^{234}_{91}\mathrm{UX}_2 \rightarrow {}^{234}_{92}\mathrm{U}$, ${}^{228}_{88}\mathrm{MsTh}_1 \rightarrow {}^{228}_{89}\mathrm{MsTh}_2 \rightarrow {}^{228}_{90}\mathrm{RdTh}$, ${}^{214}_{82}\mathrm{RaB} \rightarrow {}^{214}_{83}\mathrm{RaC} \rightarrow {}^{214}_{84}\mathrm{RaC'}$, ${}^{212}_{82}\mathrm{ThB} \rightarrow {}^{212}_{83}\mathrm{ThC} \rightarrow {}^{212}_{84}\mathrm{ThC'}$, ${}^{210}_{82}\mathrm{RaD} \rightarrow {}^{210}_{83}\mathrm{RaE} \rightarrow {}^{210}_{84}\mathrm{Po}$. Unfortunately the ground-to-ground state transitions are unobserved for many of these disintegrations, but we may make the following remarks on the basis of present information. First, the single point for $^{234}_{91}$UX$_2$ does not lie in the same Sargent band as the higher energy point for $^{234}_{90}$UX$_1$ on fig. 18. If the latter point represents the 'ground-to-ground' state transition UX$_1$→UX$_2$,† this result provides added evidence that there is some undiscovered feature of the disintegration UX$_2$→U$_{II}$ (see p. 99). Secondly, the position of the point for the unobserved ground-to-ground state transition $^{228}_{89}$MsTh$_2$→$^{228}_{90}$RdTh (Campbell, Henderson and Kyles, 1952, and unpublished) shows that the unobserved ground-to-ground state transition $^{228}_{88}$MsTh$_1$→$^{228}_{89}$MsTh$_2$ is much more highly forbidden than could be ascertained from the upper limit experimentally set for its intensity. Thirdly, a closely similar remark applies to the unobserved ground-to-ground state transition $^{214}_{82}$RaB→$^{214}_{83}$RaC, as can be seen by comparing the positions of the points representing this transition and the (observed) ground-to-ground state transition $^{214}_{83}$RaC→$^{214}_{84}$RaC′ on fig. 17. Fourthly, the points for the ground-to-ground state transitions $^{212}_{82}$ThB→$^{212}_{83}$ThC→$^{212}_{84}$ThC′ on fig. 17 lie below the line I'_a, and at distances from the allowed line O which are not very different. These points might well refer to identically forbidden transitions (the 'point' for ThB is not very accurately determined) if reason could be found for the existence of another (second-forbidden) Sargent line parallel to the line for allowed transitions. As pointed out by Feather and Richardson (1948) there is one possibility of this (one of the special cases referred to above in relation to (3·4) and (3·5)) which does not introduce serious difficulties. If the tensor form of interaction (Gamow-Teller rules) is used in calculating K_n, transitions of the type $I_i = I_f = 0$ with no change of parity are formally second-forbidden but behave in respect of the dependence of λ on W_0 as allowed transitions. We are tempted to conclude, therefore, that the spin of $^{212}_{83}$ThC in the ground state is zero and that the line OO_2 in fig. 17 represents uniquely transitions of the special type (0→0, 'no'). We return now to the last example of paired disintegrations, namely the disintegration pair

$$^{210}_{82}\mathrm{Ra\,D} \rightarrow {}^{210}_{83}\mathrm{Ra\,E} \rightarrow {}^{210}_{84}\mathrm{Ra\,F}.$$

† The argument is not affected by the fact that the nucleus $^{234}_{91}$UX$_2$ is a metastable isomer of the true ground-state nucleus $^{234}_{91}$UZ.

Even accepting a generous degree of uncertainty in our knowledge of the intensity of the ground-to-ground state disintegration Ra D→Ra E, we conclude that the two disintegrations of this pair cannot be identically forbidden (see fig. 17), unless, for the degree of forbiddenness involved, the dependence of λ on W_0 is more rapid than is represented by the Sargent lines O and I_a—and we deduce from this that the common value of $|\Delta I|$ for each disintegration is at least equal to 2. Latterly it has been general to assume that in fact $|\Delta I|=2$, and that the parity change in each case is 'no' (cf. Butt and Brodie (1951) for the most recent supporting evidence).† Then on the basis of Gamow-Teller rules the transitions are second-forbidden and (3.4) applies. With this conclusion accepted, and ρ assumed constant, we are able to draw the lines II_a, II'_a in the figure. II'_a is drawn through the point for Ra E, since $I_i > I_f$, and II_a is located higher on the diagram by 0·699 ($\log_{10} 5$) in view of the magnitude of $|\Delta I|$. We may note that empirically, at least (see below), the disintegrations $^{214}_{83}$Ra C→$^{214}_{84}$Ra C′* (1·52 MeV. excited state) and $^{212}_{83}$Th C→$^{212}_{84}$ThC′* (0·72 MeV. excited state) appear to belong to the same class as the disintegration Ra E→Ra F. Further, on the assumption that specifically structure-dependent factors in K_n are closely the same for $Z=91\pm2$ as for $Z=82\pm2$, a corresponding Sargent band ($II_a II'_a$) has been added to fig. 18, correction having been made only for the change in mean values of Z and ρ between the limits in question. It will be seen that in this way the transitions $^{239}_{93}$Np→$^{239}_{94}$Pu (ground state) and $^{226}_{89}$Ac→$^{226}_{90}$Th (ground state) are identified as probably of the class which we are discussing.

Of the 'observed' points which lie definitely below the lines II'_a on figs. 17 and 18 only two have not been referred to already. These are the points for the transitions $^{238}_{93}$Np→$^{238}_{94}$Pu* (75 keV. excited state) and $^{204}_{81}$Tl→$^{204}_{82}$Pb (ground state). Accepting the validity of our scheme in general we have to conclude that for these transitions $|\Delta I|\geqslant 3$. It must be admitted that for $^{204}_{81}$Tl this conclusion is not without its difficulties (Braid, 1951); also a general difficulty on the purely empirical level may be appreciated by reference to fig. 17. Making the most conservative assumption that the disintegration

† Doubt on this point has very recently been expressed by Petschek and Marshak (1952) who favour instead the assignment 0→0, 'yes'.

${}^{204}_{81}\text{Tl} \rightarrow {}^{204}_{82}\text{Pb}$ is of the type $3 \rightarrow 0$, the shape of the Sargent line (II'_b or III'_a, according to our notation) through the representative point is determined essentially by the factor $(\rho W_0)^4$ of (3·5) or (3·4). Such a line is drawn in the figure, and it can be appreciated at once from its intersections with II_a and II'_a that the empirical assignment of degree of forbiddenness becomes indefinite when $E_0 > 2$ MeV. If this is in fact the case, a natural limitation to the usefulness of the Sargent diagram (or for that matter of tabulated values of 'comparative half-life') is evident for heavy radioelements for which the disintegration energy is large. The effect is here very similar to the limitation expressed by (3·3), but the origin of the effect is different. Our present confusion arises from the acceptance of the transition type ($2 \rightarrow 0$, 'no') for RaE→RaF, the disintegration constant for which, in comparison with that for an allowed transition of the same energy, is nearly two orders of magnitude greater than is predicted by (3·4). This apparent blurring of the distinction between different degrees of forbiddenness need not necessarily extend to lower ranges of Z, on the other hand the condition expressed by (3·3) becomes more restrictive as Z decreases.

In turning to a survey of information concerning energies and lifetimes of 'artificially produced' β-emitters of smaller Z, we proceed to the extreme in that direction and confine attention to β-active species for which $Z < 22$. There are two reasons for this precise choice: in the first place the heaviest known mirror nuclei are the pair ${}^{41}_{20}\text{Ca}$, ${}^{41}_{21}\text{Sc}$, and secondly, over the chosen range of Z, to an entirely adequate approximation, we may write $f(0, W_0)$ instead of $f(Z, W_0)$ in (3.1)—an approximation which becomes progressively inadequate as Z increases further. That the approximation is valid implies that the effect of the electrostatic field of the nucleus on β-disintegration is practically insignificant, and it implies in particular that the distinction between positron emission and negative electron emission, which assumes increased importance at higher Z, remains all but negligible for $Z < 22$. We shall be considering positron emitters separately in the next section: here the present survey will conclude our treatment of the experimental data concerning negative electron emitters—we shall not attempt to fill in the gap $21 < Z < 80$ by detailed discussion of results. When such results were very much less extensive and generally less accurate than they are now, the attempt was made by Itoh (1940);† at the present state of knowledge it would not reveal anything of importance beyond that which we are able to extract from discussions over the two extreme ranges $Z<22$ and $Z>79$, here considered.

Fig. 19 contains the experimental points for the observed and for some 'unobserved' transitions of negative β-emitters having $Z<22$. Approached empirically, it shows that there is no allowed band which is as well-defined as those which we were able to draw in figs. 17 and 18. It would appear as a result that the specifically structural factor in K_0 (3.1) is much more variable in magnitude for the β-emitters of small Z than it is for those of high nuclear charge. If this structural factor is taken as representing the degree of conformity between the structure of the system constituted of the-parent-nucleus-with-one-neutron-changed-into-a-proton and the actual structure of the daughter nucleus (in the appropriate final state)—this degree of conformity being averaged over all the neutrons of the parent nucleus concerned—we need not be particularly surprised at this result. Indeed it has been anticipated in an earlier statement (p. 96). Recognizing the circumstances of the difficulty we can proceed to extract some numerical estimate of the effect as follows. We draw a representative allowed band OO' in fig. 19, fixing the line O by $\log_{10} f(0, W_0)+c_0$ (cf. p. 96), so as to take in as best we can the majority of allowed transitions having points on the diagram. We then compare the values of c_0, c_{82} and c_{91} used in placing the allowed bands on the three diagrams which we have constructed. We find $c_0-c_{82}=0{\cdot}72$, $c_0-c_{91}=0{\cdot}92$. This implies that the average degree of conformity as between initial and final nuclear states in allowed β-disintegrations of elements for which $Z=82\pm2$ is some five times less, and that for similar β-disintegrations of elements for which $Z=91\pm2$ some eight times less, than it is for 'typical' β-active bodies having $Z<22$. But we note by reference to fig. 19 that there are some transitions of the lightest elements, which are presumably allowed transitions, which are quite as 'unfavoured' for reasons of structure (Konopinski, 1943) as are the normal allowed transitions of the heaviest radio-

† This little-known paper is outstanding amongst earlier discussions in respect of the soundness of its approach to the problem and the significance of the conclusions reached.

elements, and we note also that three transitions (the β-disintegrations ${}^1_0n \rightarrow {}^1_1H$, ${}^3_1H \rightarrow {}^3_2He$ and ${}^6_2He \rightarrow {}^6_3Li$) are from 12 to 40 times more favoured than is represented by the arbitrarily drawn line O taken as reference line in the discussion. Of the three last-mentioned transitions two are between pairs of mirror species. A consideration of this observation will be deferred until the more numerous cases of positron disintegration of mirror species can be taken into account (p. 109), here we merely draw attention to the fact that the lie of the 'theoretical' allowed band OO' on fig. 19 is sufficiently representative of the trend of the experimental points to warrant the conclusion that the Fermi theory gives the correct dependence of λ on W_0 (or E_0) for the transitions in question. We could hardly have attached any significance to our earlier and similar conclusion (p. 96) in respect of the allowed transitions of radioelements of highest Z, if in fact it were not true in the first instance for the lightest radioelements.

Our procedure in tentatively drawing forbidden lines in fig. 19 must be the same as previously applied to fig. 17. We choose a disintegration for which detailed information regarding transition type is available, and we draw the appropriate Sargent line with the help of (3.5)† through the representative point concerned. The first disintegration to be used for this purpose in the present case is that of ${}^{14}_6C$. Here $I_i = 0$, $I_f = 1$ and we assume (even in face of the contrary predictions from some nuclear models: see Gerjuoy, 1951) that the change of parity in the disintegration is 'yes'. The transition is then first-forbidden and the Sargent line (constructed as before for ρ constant) is appropriately labelled I. We draw I' parallel to I and distant $\log_{10} 3$ below it, and have then roughly defined the first-forbidden band on the diagram. In view of the many factors involved, it is somewhat surprising how nearly completely this band covers the experimental points in the general region of the diagram which it traverses,‡ and particularly how those transitions, which on this empirical basis are grouped together, are predominantly of a single type. Five out of seven of them are transitions leading to the

† When (3.3) is not fulfilled (3.4) is inapplicable; (3.5), however, continues to apply.

‡ A little consideration will show that this result does not depend on the placing of the allowed band OO', but only on the dependence of λ on W_0 by which it is characterized.

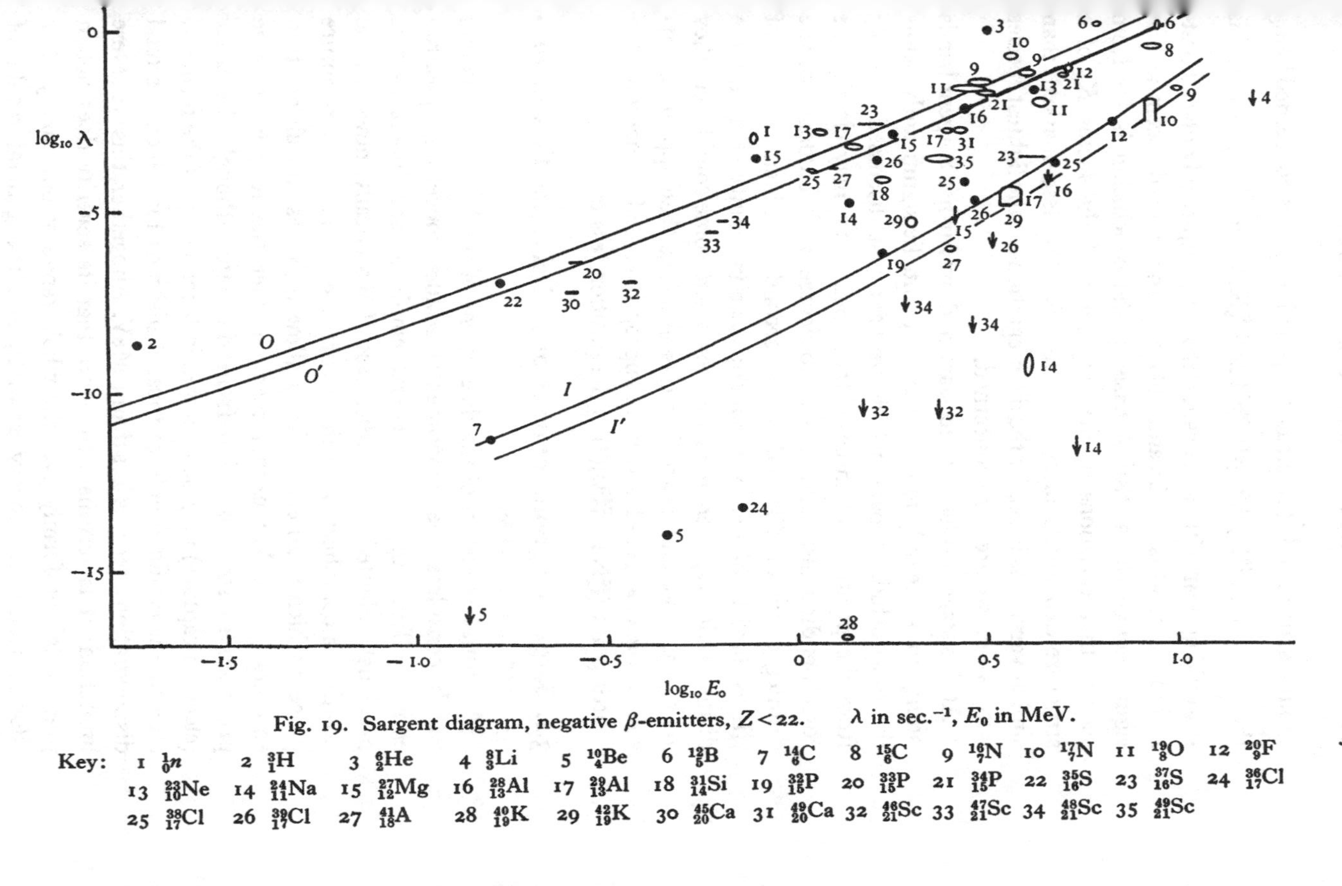

Fig. 19. Sargent diagram, negative β-emitters, $Z<22$. λ in sec.$^{-1}$, E_0 in MeV.

Key: 1 ${}^{1}_{0}n$ 2 ${}^{3}_{1}H$ 3 ${}^{6}_{2}He$ 4 ${}^{8}_{3}Li$ 5 ${}^{10}_{4}Be$ 6 ${}^{12}_{5}B$ 7 ${}^{14}_{6}C$ 8 ${}^{15}_{6}C$ 9 ${}^{16}_{7}N$ 10 ${}^{17}_{7}N$ 11 ${}^{19}_{8}O$ 12 ${}^{20}_{9}F$
13 ${}^{23}_{10}Ne$ 14 ${}^{24}_{11}Na$ 15 ${}^{27}_{12}Mg$ 16 ${}^{28}_{13}Al$ 17 ${}^{29}_{13}Al$ 18 ${}^{31}_{14}Si$ 19 ${}^{32}_{15}P$ 20 ${}^{33}_{15}P$ 21 ${}^{34}_{15}P$ 22 ${}^{35}_{16}S$ 23 ${}^{37}_{16}S$ 24 ${}^{36}_{17}Cl$
25 ${}^{38}_{17}Cl$ 26 ${}^{39}_{17}Cl$ 27 ${}^{41}_{18}A$ 28 ${}^{40}_{19}K$ 29 ${}^{42}_{19}K$ 30 ${}^{45}_{20}Ca$ 31 ${}^{49}_{20}Ca$ 32 ${}^{46}_{21}Sc$ 33 ${}^{47}_{21}Sc$ 34 ${}^{48}_{21}Sc$ 35 ${}^{49}_{21}Sc$

ground state of an $\binom{e}{e}$ nucleus, the β-active species concerned being the $\binom{o}{o}$ species $^{16}_{7}N$, $^{20}_{9}F$, $^{32}_{15}P$, $^{38}_{17}Cl$ and $^{42}_{19}K$, respectively. From this it would appear without question that our empirically placed first-forbidden band is significantly placed—and it would seem at first sight, though the statistical basis of the conclusion is far from secure, that variations in the magnitude of the factor K_1 of a structure-dependent origin are considerably less marked than similar variations in K_0 for the disintegrations of the lightest species with which we are here concerned.

At this stage it is hardly necessary to draw further Sargent bands on fig. 19, since only four 'observed' points lie significantly below the first-forbidden band which we have just discussed. Simply from the positions of three of these points we may classify the corresponding disintegrations as to degree of forbiddenness: the disintegrations of $^{10}_{4}Be$ and $^{36}_{17}Cl$ as second-forbidden and that of $^{40}_{19}K$ as third-forbidden, without apparent ambiguity. For all three disintegrations spin changes are known with some confidence; they are $0 \rightarrow 3$, $2 \rightarrow 0$ and $4 \rightarrow 0$. The classifications then appear acceptable, if the changes of parity are appropriate. Only concerning the fourth point ($^{24}_{11}Na \rightarrow ^{24}_{12}Mg^*$) is the position less clear.

3.2. Sargent diagrams for positron emission and the electron-capture process

A natural extension of the discussions of the last section is that which considers the experimental results concerning positron emission from the same general empirical standpoint as was previously adopted in the other case. Theoretically, however, the position is complicated by the occurrence of the electron-capture process which is an energetically allowed process for all positron emitters (p. 9). We cannot then consider the former of these processes in strict isolation from the latter, though the latter (electron capture) may be reviewed independently of the former if attention is confined only to those transitions for which the total disintegration energy is less than 1 MeV. Altogether, this is a severe limitation; it is fortunate therefore that its severity does not in practice fall uniformly over the whole range of our survey. For allowed transitions theory predicts, and the general results of experiment appear to confirm, that the ratio of K-capture to positron emission is less than 1/10 for elements having $Z \leqslant 10$ so long as the maximum positron energy is greater than about 600 keV., and for elements having $Z \leqslant 20$ if the maximum positron energy is greater than about 1·2 MeV. At the other end of the scale, K-capture is at least 10 times as probable as positron emission, in allowed transitions for elements with $Z \geqslant 80$, unless the energy available for capture transformation exceeds 3·0 MeV. (see Feenberg and Trigg, 1950). In the upshot, then, we can consider almost all positron-active species of low atomic number as pure positron emitters, and very nearly all capture-active bodies in the range of the classical radioelements, or of greater Z, as if they were merely capture-active—apart from any α-activity which they may exhibit. This we do in the discussions of the present section.

Fig. 20 contains the experimental points for the observed and for some 'unobserved' transitions of positron-emitters having $Z < 22$. Except for a difference in scale, it is in every way the exact counterpart of fig. 19 of the last section. For this reason, and because of the conclusion which has been stated already (p. 104) regarding the general unimportance of nuclear charge for the comparison we are making, we transfer the Sargent lines O, O', I, I' directly from the former figure to the present one. A striking result is obvious at once. Nineteen of the thirty 'observed' points on fig. 20 lie in the region above the mean allowed band OO'. These 'superallowed' points can be divided into three groups: first, fifteen points, representing ground-to-ground state† transitions of the type

$$\binom{2Z-1}{Z} \rightarrow \binom{2Z-1}{Z-1}$$

(mirror transitions); secondly, two points representing ground-to-ground state transitions $\binom{4n+2}{2n+1} \rightarrow \binom{4n+2}{2n}$; and thirdly, two points representing transitions to excited states of the respective product nuclei and of the type $\binom{4n-2}{2n} \rightarrow \binom{4n-2}{2n-1}$. In the first group all mirror transitions from ${}^{11}_{6}C \rightarrow {}^{11}_{5}B$ to ${}^{41}_{21}Sc \rightarrow {}^{41}_{20}Ca$ are represented, with the exception of ${}^{25}_{13}Al \rightarrow {}^{25}_{12}Mg$ for which energy

† There is doubt concerning this assertion only in respect of the transition ${}^{21}_{11}Na \rightarrow {}^{21}_{10}Ne$.

determinations of sufficient accuracy are not yet available, but the second and third groups are incomplete in the sense that other transitions of the same type occur which are certainly not superallowed. We shall consider the mirror transitions first, therefore, and include in our survey the two transitions of this type, ${}^1_0n \rightarrow {}^1_1\mathrm{H}$ and ${}^3_1\mathrm{H} \rightarrow {}^3_2\mathrm{He}$, left over for further discussion from the last section.

The obvious feature of transitions between ground states of mirror nuclei is that they take place between systems for which the degree of structural conformity is a maximum. The structure-dependent factor in K_0 should thus be larger for mirror transitions than for other allowed transitions of nuclei of comparable size. If mirror transitions were the only transitions which empirically were superallowed this is all that need be said; recourse to detailed theory would be unnecessary. We could take the line $\log_{10} f(0, W_0) + c$ through the points for ${}^1_0n$ and ${}^3_1\mathrm{H}$ on fig. 19 as being the ultimate allowed line for transitions for which $\Delta I = 0$, and discuss the position of all other points with reference to this line. On such a basis all the points for the mirror transitions on fig. 20 would fall slightly below the line (by amounts representing dividing factors of between 1·7 and 6 in λ), and some of this departure we could ascribe to the small effect of nuclear charge in suppressing positron emission in relation to negative electron emission of equal energy. But, as we have already mentioned, four other transitions of positron-active species are empirically superallowed on the basis of fig. 20, and the transition ${}^6_2\mathrm{He} \rightarrow {}^6_3\mathrm{Li}$ remains over for consideration from fig. 19 as the most obviously superallowed of all (factor of 40 in λ).† Again, as reference to that figure shows, a further group of four transitions of β-active species separates itself out from the rest as superallowed to a smaller degree. These are the transitions ${}^{12}_5\mathrm{B} \rightarrow {}^{12}_6\mathrm{C}^*$ (7·1 MeV.), ${}^{17}_7\mathrm{N} \rightarrow {}^{17}_8\mathrm{O}^*$ (5·0 MeV.), ${}^{23}_{10}\mathrm{Ne} \rightarrow {}^{23}_{11}\mathrm{Na}^*$ (3·0 MeV.) and ${}^{27}_{12}\mathrm{Mg} \rightarrow {}^{27}_{13}\mathrm{Al}^*$ (1·85 MeV.): concerning them we shall only remark that a high excited state of the product nucleus is involved in each case, and that three of the four parent species are of the type $\binom{2n+3}{n}$.

† This transition provides a clear example of the type $I_i = 0$, $I_f = 1$. Presumably, therefore, it owes its unique position to the operation of the Marshak weight factor (p. 98) (see also Moszkowski, 1951).

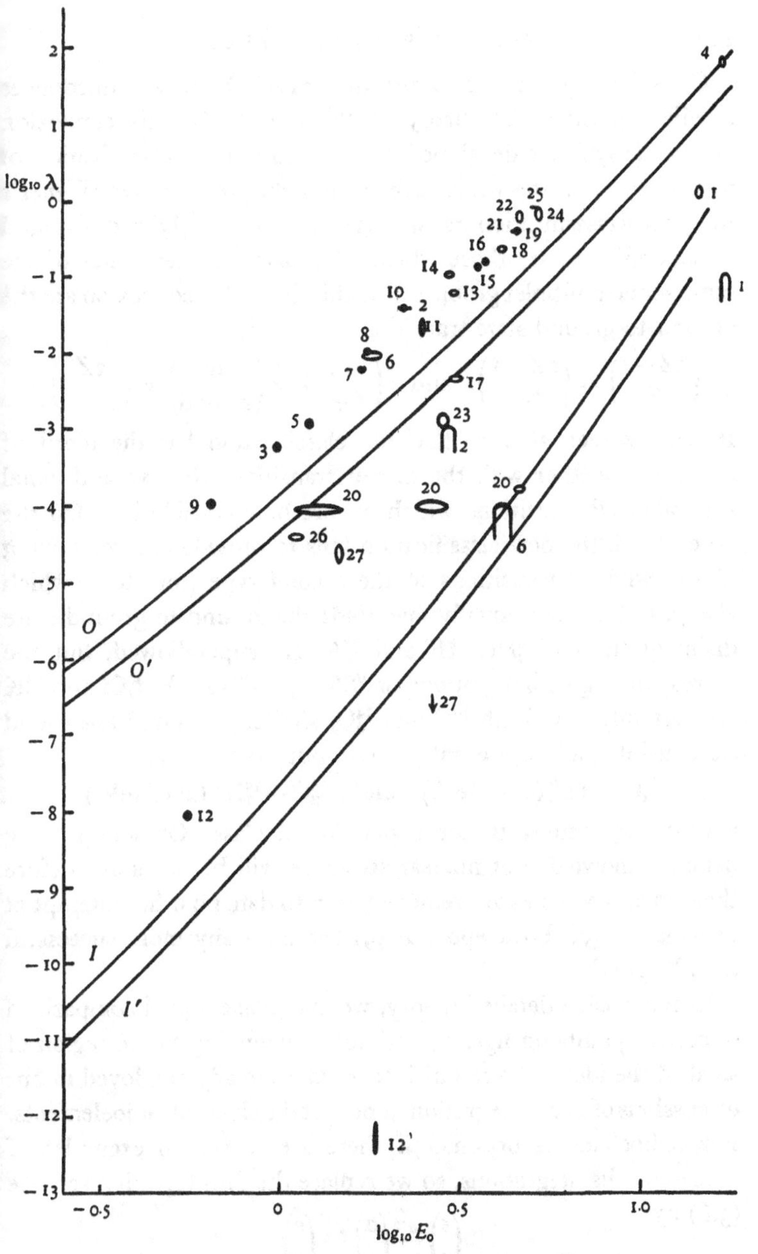

Fig. 20. Sargent diagram, positron emitters, $Z < 22$. λ in sec.$^{-1}$, E_0 in MeV.

Key:

1	${}^{8}_{5}B$	2	${}^{10}_{6}C$	3	${}^{11}_{6}C$	4	${}^{12}_{7}N$	5	${}^{13}_{7}N$	6	${}^{14}_{8}O$
7	${}^{15}_{8}O$	8	${}^{17}_{9}F$	9	${}^{18}_{9}F$	10	${}^{19}_{10}Ne$	11	${}^{21}_{11}Na$	12	${}^{22}_{11}Na$
13	${}^{23}_{12}Mg$	14	${}^{26}_{13}Al$	15	${}^{27}_{14}Si$	16	${}^{29}_{15}P$	17	${}^{30}_{15}P$	18	${}^{31}_{16}S$
19	${}^{33}_{17}Cl$	20	${}^{34}_{17}Cl$	21	${}^{35}_{18}A$	22	${}^{37}_{19}K$	23	${}^{38}_{19}K$	24	${}^{39}_{20}Ca$
25	${}^{41}_{21}Sc$	26	${}^{43}_{21}Sc$	27	${}^{44}_{21}Sc$						

Consideration of the other five cases, however, introduces detailed questions of theory at the outset. In this connexion Wigner (1939) has developed a 'super-multiplet' classification of nuclear energy states which is relevant to the problem. On Wigner's theoretical classification transitions which empirically are designated as superallowed are those which take place between states of the same super-multiplet group.† Possibilities in this connexion are the ground-to-ground state transitions

$$\binom{2Z-1}{Z} \rightleftarrows \binom{2Z-1}{Z-1} \quad \text{and} \quad \binom{2Z}{Z+1} \rightleftarrows \binom{2Z}{Z,\ \text{odd}} \rightleftarrows \binom{2Z}{Z-1}.$$

It will be seen at once that the classification has the merit of grouping together with the mirror transitions the five additional superallowed transitions which we are now considering. On the other hand, the new classification fails to provide any convincing reason why all transitions of the second type (for all of which $A=4n+2$) are not equally favoured: the ground-to-ground state disintegrations of $^{6}_{2}$He, $^{18}_{9}$F and $^{26}_{13}$Al are superallowed, but the corresponding disintegrations of $^{10}_{4}$Be, $^{14}_{6}$C, $^{22}_{11}$Na, $^{30}_{15}$P, $^{34}_{17}$Cl and $^{38}_{19}$K are certainly not. With $^{10}_{6}$C and $^{14}_{8}$O, similarly, ground-to-ground state disintegrations are unfavoured, but the transitions

$$^{10}_{6}\text{C} \rightarrow {}^{10}_{5}\text{B}^{*}\,(0{\cdot}7\ \text{MeV.}) \quad \text{and} \quad {}^{14}_{8}\text{O} \rightarrow {}^{14}_{7}\text{N}^{*}\,(2{\cdot}3\ \text{MeV.})$$

appear surprisingly in the superallowed class. Obviously more detailed knowledge of nuclear structure will be necessary before these inconsistencies are removed. Up to date no other attempt at explanation (cf. Dzhelepov, 1949) has been any more successful than Wigner's.

Leaving aside detailed theory, we may make a final comparison of certain points on figs. 19 and 20 by extending to the region of small Z the idea of paired disintegrations already employed in our discussions of β-disintegrations amongst the classical radioelements. For radioelements of small Z there are as yet no examples of successive disintegrations, so we replace the disintegration scheme (3.6) by

$$\binom{e}{e} \xrightarrow{\beta^-} \binom{o}{o} \xleftarrow{\beta^+} \binom{e}{e}$$

† A super-multiplet group in the Wigner sense is not a set of states of a particular nucleus $\binom{A}{Z}$; it is a set, having specified symmetry properties, amongst an isobaric 'pleiad' A constant.

as defining the pairs under discussion. For any such pair, both active nuclei having $I=0$ and even parity according to accepted ideas, we expect the transitions of each to a common state of the stable $\binom{o}{o}$ nucleus to be identically forbidden. We have just two pairs to consider: $^{10}_{4}Be$ and $^{10}_{6}C$, and $^{14}_{6}C$ and $^{14}_{8}O$. For the first, the ground-to-ground state disintegration of $^{10}_{4}Be$ has been classified as second-forbidden, that of $^{10}_{6}C$ is unobserved; the disintegration $^{10}_{4}Be \rightarrow ^{10}_{5}B^{*}$ (0·42 MeV.) is at least second-forbidden, the corresponding disintegration of $^{10}_{6}C$ is also unobserved; the disintegration $^{10}_{6}C \rightarrow ^{10}_{5}B^{*}$ (0·7 MeV.) is superallowed, the corresponding disintegration of $^{10}_{4}Be$ is energetically impossible. For the second, the ground-to-ground state disintegration of $^{14}_{6}C$ has been classified as first-forbidden, that of $^{14}_{8}O$ is unobserved and the position of the point on fig. 20 indicates that it may still be first-forbidden (assuming a relative intensity of about 0·01); the disintegration $^{14}_{8}O \rightarrow ^{14}_{7}N^{*}$ (2·3 MeV.) is superallowed, the corresponding disintegration of $^{14}_{6}C$ is energetically impossible. There is no absolute confirmation of prediction here, but also nothing clearly contrary to expectations.

We turn now to a consideration of the λ/E_0 relation for the electron-capture process. The overriding difficulty here is the determination of E_0, the energy of transition. The whole of this energy is carried away by the emitted neutrino, and is thus unobservable. If the transition is a simple transition between ground states, the neutrino is the only nuclear radiation emitted; in any other case the determination of branching ratios depends essentially on an absolute determination of the intensities of the nuclear γ-rays emitted in the reorganization of the product nucleus. In such a case partial disintegration constants, as well as transition energies, are difficult to evaluate experimentally. In view of all these considerations it is fortunate that just with the heaviest capture-active bodies, which for other reasons (p. 109) are the most significant to study, the possibility arises of an accurate estimation of ground-to-ground state transition energy by indirect means. The method employed is that of the closed decay cycle. In such a cycle, if all but one of the ground-to-ground state transition energies are known, the remaining transition energy can be determined by use of the conservation law. In the region of the classical radioelements accurately determined α-particle energies provide a background of exact information, and in any closed cycle of four transitions furnish the transition energies for two of the three 'knowns'. The third known energy may be the experimentally determined total energy of disintegration of a negative β-active body, or may be the ground-to-ground state transition energy of a capture-active body which has already been determined by the closure of another cycle. In these calculations, of course, allowance must be made for the ionization energy of the residual atom (K-ionization energies of the order of 0·1 MeV. in this range of Z).†

Data obtained in the way which has just been described have been reviewed by Thompson (1949) and by the writer (Feather, 1952*b*). The chief difference in approach as between these two discussions lies in the greater attention paid to transitions other than between ground states in the later survey. The considerable advance in experimental knowledge of accompanying γ-radiation during the intervening period invited this more detailed treatment; its neglect, or impracticability, in the earlier discussion appears to have seriously vitiated the conclusions drawn. Readers are referred to the original paper (Feather, loc. cit.) for the sifting of evidence, here we merely state that in the upshot only twelve points could be plotted with any assurance (and not all of these with complete confidence, see legend of fig. 21) on a Sargent diagram, having reference to the range of atomic numbers $89 \leqslant Z \leqslant 98$. This diagram is reproduced here as fig. 21. Before we proceed to its discussion we refer briefly to the predictions of theory.

For allowed capture transitions all currently considered versions of the Fermi theory give essentially the same result. To a sufficiently good approximation (having regard to the nature of the experimental results) we may write

$$\lambda_0 = C_0 Z^3 (E_K^2 + \mu_0 E_L^2), \tag{3·7}$$

where, apart from explicitly known factors, C_0 is identical with K_0 of (3·1) (as referred, of course, to the same pair of initial and final states), E_K and E_L denote the transition energies for capture of K

† A preliminary report of a general method of determining E_0, by a study of the weak 'bremsstrahlung' emitted as the result of electron capture, has recently been given by Maeder and Preiswerk (1951).

and L_{I} electrons, respectively, and μ_0 is a quantity of the order of magnitude 0·17 for the range of Z with which we are here concerned (Rose and Jackson, 1949). Necessarily $E_L > E_K$, and clearly only positive values of these transition energies have any meaning in the equation.

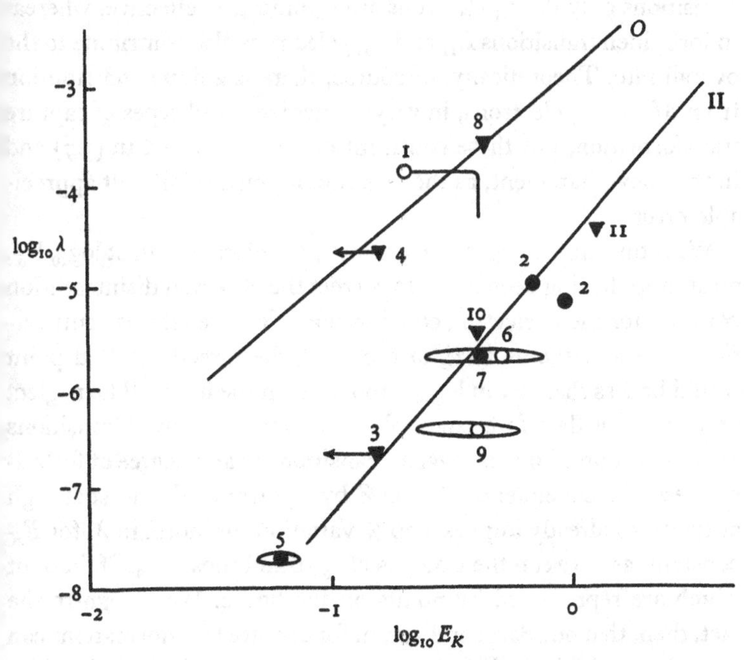

Fig. 21. Sargent diagram, capture-active species, $89 \leqslant Z \leqslant 98$. λ in sec.$^{-1}$, E_K in MeV., ● more reliable, ▼ reliable, ○ less reliable.

Key: 1 $^{244}_{98}$Cf 2 $^{239}_{95}$Am 3 $^{237}_{94}$Pu 4 $^{234}_{94}$Pu 5 $^{235}_{93}$Np 6 $^{234}_{93}$Np
7 $^{231}_{92}$U 8 $^{228}_{92}$U 9 $^{230}_{91}$Pa 10 $^{229}_{91}$Pa 11 $^{224}_{89}$Ac

For forbidden transitions the position is not so simple. We can no longer retain the factor $Z^3(E_K^2 + \mu_0 E_L^2)$ of (3.7), as the factor $f(Z, W_0)$ was taken over from (3.1) into (3.2) without undue error, concentrating all the factors of forbiddenness in K_n, but we can still say, in general, that a distinction such as that expressed in (3.3) to (3.5) is relevant (and here the distinction between Fermi and Gamow-Teller rules reappears), and we can say approximately, and in general, that, where a factor of forbiddenness of the order of $(\rho W_0)^{2n}$ would appear in relation to β-disintegration, corresponding factors $(\rho E_K)^{2n}$ and $(\rho E_L)^{2n}$ appear multiplying the terms expressing the contributions to λ from K- and L-electron capture respectively. But the contribution from L-electron capture increases as the order of forbiddenness increases (to an extent dependent on the type of rules assumed), just because in allowed transitions only the L_I electrons are significantly effective, whereas in forbidden transitions L_{II} and L_{III} electrons also contribute to the overall rate. Theoretically, of course, there is a finite contribution from $M, N, \ldots,$ electrons, in varying degree, to all types of capture transformation, but these contributions are neglected in (3·7) and in the above statement, as they can be in practice without appreciable error.

We return now to fig. 21. Here $\log_{10} \lambda$ is plotted against $\log_{10} E_K$, no attempt having been made to correct the observed disintegration constant for the L-electron contribution. Yet, even if this contribution were as large as 20% of the total, the directly plotted point would be less than 0·1 in $\log_{10} \lambda$ too high on the figure. The neglect of the effect is therefore excusable, at least when allowed transitions are in question. And, as regards transitions of any degree of forbiddenness, the dependence of λ on Z by a power of Z at least as high as the third already implies a 30% variation (or more) in λ, for E_K constant, as between the isotopes of $_{89}$Ac and those of $_{98}$Cf, both of which are represented by points on the figure. We recognize the fact, then, that our Sargent diagram for capture transformations can at the best exhibit the broad features of the theoretical relationship, and we concentrate attention on an attempt to elicit these features. Proceeding empirically, we accept the point for the capture transformation of $^{228}_{92}$U as the best-attested point for a transition which can provisionally be classified as allowed, and we draw the line O through that point with the 'theoretical' slope 2 as given by the first term in (3·7). We note that within the indicated uncertainty of present experimental knowledge the points for $^{244}_{98}$Cf and $^{234}_{94}$Pu also lie on this line, and that no point lies significantly above it. In somewhat finer detail we note that the three capture-active species $^{228}_{92}$U, $^{234}_{94}$Pu and $^{244}_{98}$Cf are the only $\begin{pmatrix} e \\ e \end{pmatrix}$ species represented on the diagram, and we tend to conclude that if the position of our allowed line is significant to $\pm 0{\cdot}3$ in $\log_{10} \lambda$ then it is probably true to say that it is more correctly labelled O than O', according to our normal convention (cf. p. 98). As regards the absolute position of this line, we may use the method of comparative half lives (footnote, p. 100) to compare the placings of the lines O in figs. 21 and 17 (β-disintegrations, $80 \leqslant Z \leqslant 84$). Taking the effective value of Z for fig. 21 as 93 and that for fig. 17 as 82, we have $\log_{10}(ft) = 5{\cdot}2$ for the allowed line O in fig. 17 and $\log_{10}(ft) = 4{\cdot}6$ for the tentatively identified allowed line in fig. 21. The agreement (to within a factor of 4 in λ) is as close as could be expected. We can reasonably conclude, then, that our initial assumption that $^{228}_{92}U$ transforms by an allowed capture-process is in fact correct.

When the line O has been drawn in fig. 21, all the remaining nine points can be accounted for, almost within the assigned limits of experimental uncertainty, by the straight line II. The slope of this line is considerably greater than 2—as befits the line for a forbidden transition—and it has been labelled II rather than I on the assumption that it certainly does not represent first-forbidden transitions of the type ($\Delta I = 0, \pm 1$; 'yes'). That it might belong to the transition type ($\Delta I = \pm 2$, 'yes'), also first forbidden, and to which a result corresponding to (3.5) rather than (3.4) would apply, remains a possibility, but we have at present no means of deciding between this and the suggestion implicit in the labelled figure. In principle the method of the closed cycle can obviously be used in respect of conservation of spin and parity, as it has already been used in respect of conservation of energy, but the information available is as yet insufficient for this purpose so far as any of the transitions represented by points on II, fig. 21, are concerned.

It has already been stated that a fairly severe scrutiny of experimental evidence was carried out before points were accepted as sufficiently well authenticated for plotting in fig. 21. In this process the capture transformations of some fifteen species in the range $89 \leqslant Z \leqslant 98$ were disregarded, although a good estimate of ground-to-ground state transition energy was available in each case. It should here be said that if we make the arbitrary assumption that each of these species transforms in a single mode without excitation of the residual nucleus we have a further rough means of testing our Sargent diagram for plausibility. Hypothetical points may be plotted for the previously disregarded species and a survey made of the distribution of these points on the diagram. It is satisfactory that none of the hypothetical points so plotted lies above the allowed line O.

There is just one further case of capture transformation which can profitably be discussed here, though it falls completely outside our self-imposed region of survey, that of the classical radio-elements (p. 109). The transformation ${}^{7}_{4}Be \rightarrow {}^{7}_{3}Li$, between mirror species, is energetically impossible by positron emission, the available energy being accurately known from a determination of the threshold energy of the reaction ${}^{7}_{3}Li(p, n){}^{7}_{4}Be$. As a capture transformation it has been thoroughly studied, and competing modes (ground-to-ground state 89%, ground-to-excited state 11%) have been established. The $\log_{10}(ft)$ values for these modes are 3·36 and 3·56, respectively (Feingold, 1951). On this basis each mode is empirically superallowed—for the $\log_{10}(ft)$ values for the positron disintegrations of the $\binom{2Z-1}{Z}$ species of fig. 20 fall in just this range (e.g. values of 3·59 and 3·40 for ${}^{11}_{6}C$ and ${}^{41}_{21}Sc$, the extreme members of the series). An additional item of regularity is thus exposed, and the interesting conclusion emerges that the ground state of the residual nucleus ${}^{7}_{3}Li$, and the excited state of 480 keV. energy, belong to the same super-multiplet on the Wigner classification.

3.3. Disintegration energies and relative stability rules

Measurement of β-particle, and where necessary of associated γ-ray, energies makes possible the determination of differences of mass between neighbouring isobars. More specifically (see § 1.2), if $T_{\beta}^{-}\binom{A}{Z}$ is the ground-to-ground state transition energy in the negative β-particle disintegration $\binom{A}{Z} \rightarrow \binom{A}{Z+1}$ we have

$$T_{\beta}^{-}\binom{A}{Z} = \left[N\binom{A}{Z} - N\binom{A}{Z+1} - m\right]c^2$$
$$+ [W(Z+1) - W(Z) - W_v(Z+1)]. \quad (3.8)$$

Here $N\binom{A}{Z}$ is the mass of the nucleus $\binom{A}{Z}$ in the ground state, mc^2 is the rest-energy of the electron, $W(Z)$ is the total energy of binding of the atomic electrons in the neutral atom $\binom{A}{Z}$, and $W_v(Z)$ is energy of binding of the outermost (valency) electron in the same atom. This result implies (for the sake of definiteness) that the disintegrating atom is originally neutral, and that the whole of the energy of atomic reorganization is available for the outgoing β-particle. We may re-write (3.8) in the form

$$\left[N\binom{A}{Z} - N\binom{A}{Z+1}\right]c^2 = T_\beta^-\binom{A}{Z} + mc^2 - \epsilon'\binom{Z+1}{Z}. \quad (3.9)$$

For the positron disintegration $\binom{A}{Z} \rightarrow \binom{A}{Z-1}$ the corresponding result is

$$\left[N\binom{A}{Z} - N\binom{A}{Z-1}\right]c^2 = T_\beta^+\binom{A}{Z} + mc^2 + \epsilon\binom{Z}{Z-1}, \quad (3.10)$$

the only difference between the functions ϵ' and ϵ being that the term corresponding to $W_v(Z+1)$ does not appear in ϵ. To a very high degree of accuracy, therefore,

$$T_\beta^-\binom{A}{Z} + T_\beta^+\binom{A}{Z+1} = -2mc^2. \quad (3.11)$$

In the first part of this section we shall use equations (3.9) to (3.11) in a survey of the results of experiment; in the second part certain empirical generalizations will be examined.

From the whole range of experimental results by far the most coherent set of data refers to the mirror transitions of the positron-active species $^{11}_{6}C$ to $^{41}_{21}Sc$. For these elements the term $\epsilon\binom{Z}{Z-1}$ in (3.10) is completely negligible, thus the difference in mass is given by $m + T_\beta^+/c^2$, alone. In theory this difference in mass between mirror nuclei arises merely from the difference in electrostatic energy and the difference in mass between neutron and proton—so long as the exact equivalence of the specifically nuclear forces between proton and proton, and neutron and neutron, is assumed. Again, assuming that the protons in the nucleus are randomly distributed throughout a sphere of radius $r_0A^{\frac{1}{3}}$, the electrostatic energy of the nucleus $\binom{A}{Z}$ is $\frac{3}{5}\frac{e^2}{r_0}\frac{Z(Z-1)}{A^{\frac{1}{3}}}$, and, if $T_\beta^-\binom{1}{0}$ is the β-disintegration energy of the neutron, we have

$$\left[N\binom{A}{Z}-N\binom{A}{Z-1}\right]c^2=\frac{6}{5}\frac{e^2}{r_0}\frac{Z-1}{A^{\frac{1}{3}}}-mc^2-T_\beta^-\binom{1}{0}. \quad (3.12)$$

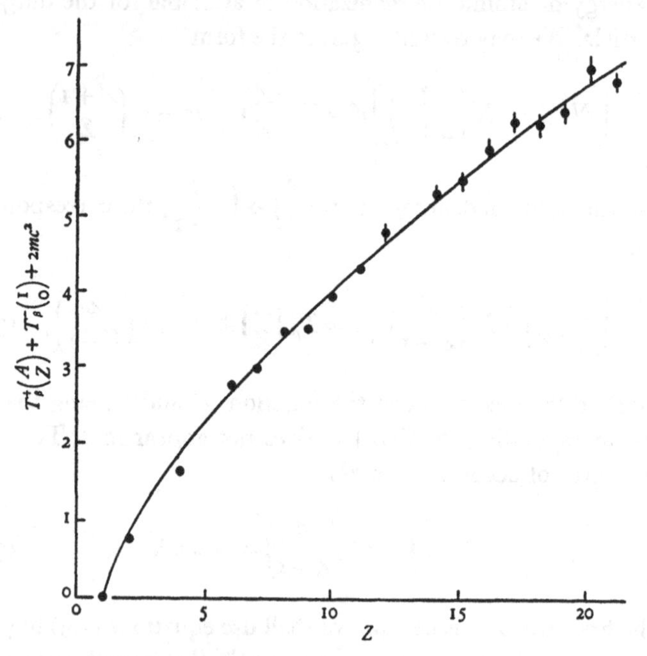

Fig. 22. Energy of positron disintegration of 'mirror' species as a function of A (<42).

$$T_\beta^+\binom{A}{Z}+T_\beta^-\binom{1}{0}+2mc^2 \text{ in MeV.}$$

From (3.12) and (3.10), neglecting $\epsilon\binom{Z}{Z-1}$, we obtain

$$\frac{6}{5}\frac{e^2}{r_0}\frac{Z-1}{A^{\frac{1}{3}}}=T_\beta^+\binom{A}{Z}+T_\beta^-\binom{1}{0}+2mc^2. \quad (3.13)$$

In fig. 22 plotted points represent the values of the right-hand member of (3.13) for the species $^{11}_{6}$C to $^{41}_{21}$Sc, and the smooth curve values obtained from the left-hand member of the same equation with $r_0=1{\cdot}44\times10^{-13}$ cm. It is remarkable how closely the experimental results are reproduced using such a simple and essentially classical' model of the nucleus, and how nearly the value of r_0 deduced from this model as applied to the light nuclei here concerned agrees with the value of the fundamental radius deduced from the experimental results regarding the α-disintegration of the heaviest nuclei on the basis of the Gamow theory (see § 2.2). For a discussion of the electrostatic energy in terms of more refined nuclear models the reader is referred to a review by Wigner and Feenberg (1941).

The close agreement between theory and experiment which has just been noted extends to values of Z even less than 6. Points can be plotted on fig. 22 for the species $^{7}_{4}$Be, $^{3}_{2}$He and $^{1}_{1}$H. All these species are stable against positron disintegration, but the (negative) values of T^{+}_{β} appropriate to them can be deduced, in the first case from the energy of capture transformation of the species in question (p. 118), and in the other two through (3.11) from the β-disintegration energies of $^{3}_{1}$H and $^{1}_{0}n$. In the last case, of course, (3.11) implies that the electrostatic energy difference is identically zero—a self-evident result, as the expression $\frac{6}{5}\frac{e^2}{r_0}\frac{Z-1}{A^{\frac{1}{3}}}$ indicates—but for the others there is no such *a priori* reason why experiment and theoretical calculation should agree. Yet the additional points (for $^{7}_{4}$Be and $^{3}_{2}$He) on fig. 22 fall on the smooth curve almost as well as those representing the positron disintegrations of the heavier species.

We have just been discussing the variation with Z of the total disintegration energy for positron emission, $T^{+}_{\beta}\binom{A}{Z}$, of species having the same isotopic number, $A-2Z=-1$. On the basis of (3.11) this variation (with change of sign) is essentially that of the total disintegration energy for negative β-emission of species for which the isotopic number is $+1$. The results elicited by the discussion suggest that it would be profitable to extend our survey to other series of species for which $A-2Z$ is constant. Such an extended survey was first made in detail by Saha and Saha (1946), but the information now available is much more complete than that with which these authors had to work, and regularities—or irregularities—of significance are thus more likely to emerge. In principle, of course, the survey could be made exhaustive,† each

† A complete survey, for $A-2Z$ odd, has recently been published by Suess and Jensen (1951).

β-active species being considered according to its isotopic number, but as in the last two sections we shall make a selection representative of the theoretically interesting regions of Z, and omit the rest.

Fig. 23 contains the experimental information for the light species for which the Sargent diagrams of figs. 19 and 20 were drawn. Here the curve of fig. 22 reappears, inverted and with change of

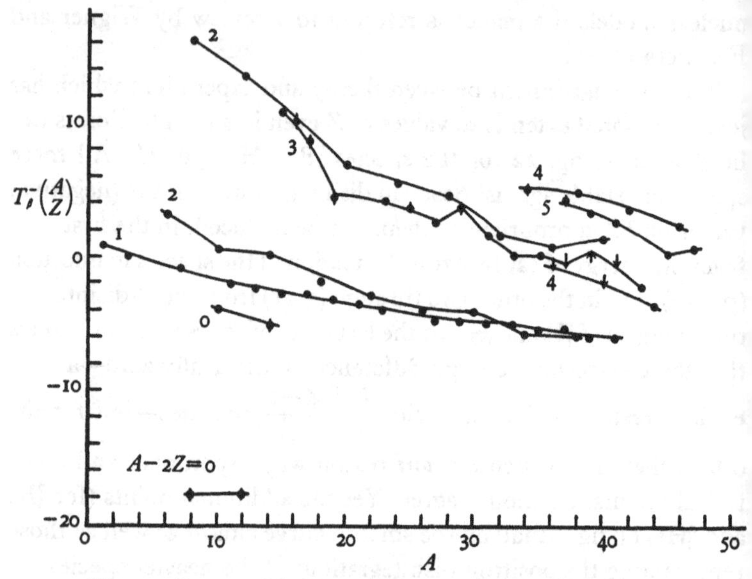

Fig. 23. Energy of β-disintegration as a function of $A(<48)$, for $A-2Z$ constant.

$T_{\beta}^{-}\binom{A}{Z}$ in MeV.

origin, as the curve for species of isotopic number 1, and the other curves represent the total energies of negative β-disintegration of species for which $A-2Z=0$, 2, 3, 4 and 5, respectively. A very obvious regularity emerges at first glance: the curves for the even isotopic numbers are double, those for the odd isotopic numbers to all appearances single. This result clearly derives from the fact that species of even A (that is, even isotopic number) are of two parity types characterized by widely different energies of binding, whilst those of odd A (or odd isotopic number) though in theory distinguishable as to parity type, are effectively of a single class so far as binding energy is concerned. For even A, species of the type $\binom{e}{e}$

are tightly bound, those of the type $\binom{o}{o}$ are loosely bound. For even A, therefore, β-disintegrations represented as $\binom{o}{o} \rightarrow \binom{e}{e}$ are relatively energetic, those represented as $\binom{e}{e} \rightarrow \binom{o}{o}$ are very much less energetic—or frequently energetically disallowed, as the curves of fig. 23 show. For odd A, on the other hand, the two types of disintegration $\binom{e}{o} \rightarrow \binom{o}{e}$ and $\binom{o}{e} \rightarrow \binom{e}{o}$ are on an equal footing energetically, at least so long as the equality of the specifically nuclear forces between proton and proton, and neutron and neutron, is assumed. Even if a small degree of splitting of the odd-A curves of T_{β}^{-} against A is to be anticipated for higher values of the isotopic number, it should be entirely negligible on this basis for the small values of $A-2Z$ represented in fig. 23.

As regards detailed features, the curve for $A-2Z=3$ in fig. 23 shows most of interest. There is the sudden drop in T_{β}^{-} between ${}^{17}_{7}\mathrm{N}$ and ${}^{19}_{8}\mathrm{O}$, and there are the 'maxima' at ${}^{29}_{13}\mathrm{Al}$ and ${}^{39}_{18}\mathrm{A}$. The sudden drop is clearly explicable in terms of the tightly bound character of a shell of 8 protons. We assume that this complete shell is present in the final nucleus in the disintegration ${}^{17}_{7}\mathrm{N} \rightarrow {}^{17}_{8}\mathrm{O}$, and in the initial nucleus in the considerably less energetic ${}^{19}_{8}\mathrm{O} \rightarrow {}^{19}_{9}\mathrm{F}$. In a similar way the 'maximum' at ${}^{39}_{18}\mathrm{A}$ is explicable in terms of a closed shell of 20 neutrons present in the product nucleus ${}^{39}_{19}\mathrm{K}$ (as also in the parent nucleus in the energetically disallowed transition ${}^{37}_{17}\mathrm{Cl} \rightarrow {}^{37}_{18}\mathrm{A}$). Only the 'maximum' at ${}^{29}_{13}\mathrm{Al}$ is rather surprising. On the assumption that the experimental results are unassailable, it has been suggested (Seidlitz, Bleuler and Tendam, 1949) that a shell of 14 protons is tightly bound. But it is at least equally certain that a shell of 14 neutrons shows no such character: the disintegration ${}^{25}_{11}\mathrm{Na} \rightarrow {}^{25}_{12}\mathrm{Mg}$, which might be thought to be of low energy, is normally energetic, and the disintegration ${}^{27}_{12}\mathrm{Mg} \rightarrow {}^{27}_{13}\mathrm{Al}$, for which a high energy might have been predicted, is of low rather than of high energy release. In the upshot it is somewhat disconcerting to find protons and neutrons appearing to 'behave' differently at nucleon number 14, when seemingly they behave so similarly otherwise. On the curve for $A-2Z=2$ (Z odd) the 'maximum' at ${}^{40}_{19}\mathrm{K}$ is just what would be expected in view of constitution of the final nucleus ($N=20$, $Z=20$), and the 'minimum' at ${}^{45}_{20}\mathrm{Ca}$ on the curve $A-2Z=5$ is in like measure to be correlated with the existence of a closed shell of 20 protons in the initial nucleus.

Experimental data for the sequences $A-2Z=11$, 13 and 15 in fig. 24, and for the sequences $A-2Z=23$, 25 and 27 in fig. 25, are in each case plotted against neutron number N. In fig. 26 data for the sequences $A-2Z=17$, 19 and 21 are similarly plotted against proton number Z. In this way figs. 24 and 25 exhibit the variation of $T_{\bar{\beta}}$ with N in the regions $N \sim 50$, $N \sim 82$, and fig. 26 shows the variation of $T_{\bar{\beta}}$ with Z when $Z \sim 50$. The information in these three figures may be thought of as representing the experimental data—much extended and critically revised—first presented so as to show the effect of shell-closure by Suess (1951). Suess published curves for $A-2Z=13$ and 25 only, and did not include information available merely in the form of inequalities (such as the 'qualitative' information that a species is β-stable by greater than a specified energy, just because γ-radiation of that energy is observed in the K-capture disintegration of the neighbouring isobar of higher Z). Following our previous practice, we include 'unobserved' points (represented by arrows in figs. 23–26) whenever the qualitative information so recorded helps to fill out the general picture which the 'observed' points establish.

Figs. 24 and 25 exhibit a remarkable degree of similarity. We shall discuss first the similarity in the curves for $A-2Z=11$ and $A-2Z=23$, the last plotted points on which are those for the 'magic' neutron numbers $N=50$, $N=82$, respectively. These curves provide little or no evidence for any systematic difference in energy as between disintegrations of the two types $\begin{pmatrix} o \\ e \end{pmatrix} \rightarrow \begin{pmatrix} e \\ o \end{pmatrix}$ and $\begin{pmatrix} e \\ o \end{pmatrix} \rightarrow \begin{pmatrix} o \\ e \end{pmatrix}$. In each case the curve is practically linear, and the mean slopes (0·45 and 0·40 MeV. per neutron-proton pair added to the parent nucleus) are not very different. From (3.9) we have

$$T_{\bar{\beta}}\begin{pmatrix} A \\ Z \end{pmatrix} - T_{\bar{\beta}}\begin{pmatrix} A+2 \\ Z+1 \end{pmatrix}$$
$$= \left[N\begin{pmatrix} A+2 \\ Z+2 \end{pmatrix} - N\begin{pmatrix} A \\ Z+1 \end{pmatrix} \right] c^2 - \left[N\begin{pmatrix} A+2 \\ Z+1 \end{pmatrix} - N\begin{pmatrix} A \\ Z \end{pmatrix} \right] c^2$$
$$+ \epsilon'\begin{pmatrix} Z+1 \\ Z \end{pmatrix} - \epsilon'\begin{pmatrix} Z+2 \\ Z+1 \end{pmatrix}, \quad (3.14)$$

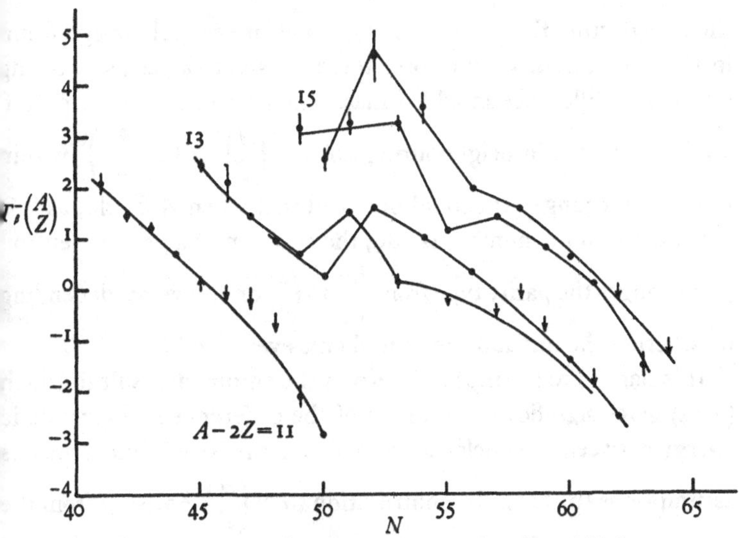

Fig. 24. Energy of β-disintegration as a function of N, for $A-2Z=11$, 13 and 15.
$T_{\beta}^{-}\binom{A}{Z}$ in MeV.

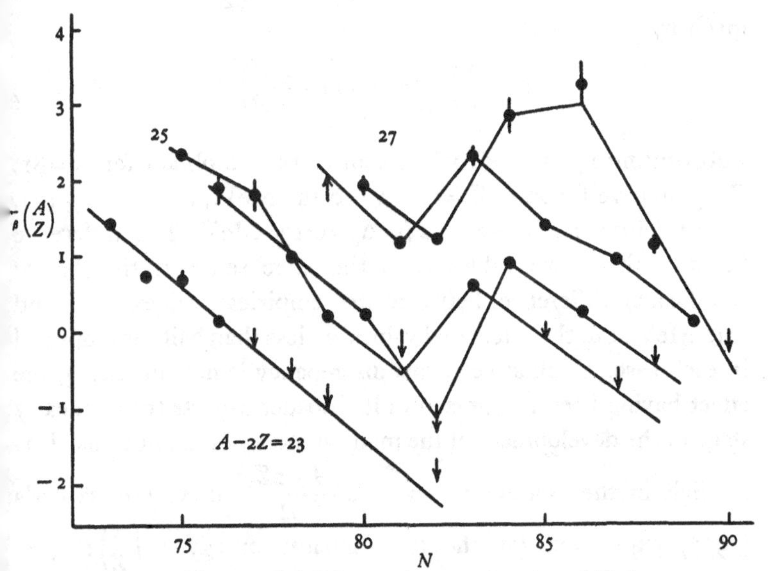

Fig. 25. Energy of β-disintegration as a function of N, for $A-2Z=23$, 25 and 27.
$T_{\beta}^{-}\binom{A}{Z}$ in MeV.

thus, neglecting the last two terms, which are entirely insignificant in this connexion, we interpret these constant slopes as implying a constant difference in added mass when a neutron and a proton are incorporated in neighbouring isobars, $\binom{A}{Z}$ and $\binom{A}{Z+1}$, within the relevant range. It should be noted that when A is odd, as it is when the isotopic number is odd, the addition of a proton-neutron pair changes the parity type from $\binom{o}{e}$ to $\binom{e}{o}$ or vice versa, depending on whether the initial proton number is even or odd.

It is natural to attempt to interpret the empirical result to which (3·14) gives significance in terms of the difference in electrostatic energy between the nuclei concerned. On the basis of our previous assumptions (p. 119), the contribution to $N\binom{A}{Z}c^2$ arising from the electrostatic energy of the nuclear protons is $\frac{3}{5}\frac{e^2}{r_0}\frac{Z(Z-1)}{A^{\frac{1}{3}}}$, thus the electrostatic contribution to $T_{\beta}^{-}\binom{A}{Z}-T_{\beta}^{-}\binom{A+2}{Z+1}$ is very closely given by

$$\Delta_e = \frac{6}{5}\frac{e^2}{r_0}\frac{1}{(A+2)^{\frac{1}{3}}}\left(1-\frac{2}{3}\frac{Z}{A}\right). \qquad (3\cdot 15)$$

Substituting $r_0 = 1{\cdot}44\times 10^{-13}$ cm. in (3·15) we obtain for $A=81$, $Z=35$ (curve for $A-2Z=11$), $\Delta_e=0{\cdot}196$ MeV., and for $A=131$, $Z=54$ (curve for $A-2Z=23$), $\Delta_e=0{\cdot}170$ MeV. The difference between these two values is in the same sense as the (barely significant) difference between the empirical values 0·45 and 0·40 MeV., but the calculated values are less than half the empirical in each case. Qualitatively, this discrepancy is not surprising, the effect having been recognized in its broader aspects from an early stage in the development of the modern theory of the nucleus. It is implicit in the 'symmetry term' $\beta\frac{(A-2Z)^2}{A}$ in v. Weizsäcker's (1935) expression for the total nuclear energy $N\binom{A}{Z}c^2$ (see Feenberg, 1947). According to Pryce (1950), recent estimates of the constant β fall within the range $20\pm 0{\cdot}5$ MeV. With this assumption, then, we calculate the contribution of the symmetry energy to

$T_{\overline{\beta}}\binom{A}{Z} - T_{\overline{\beta}}\binom{A+2}{Z+1}$. We have formally

$$\Delta_s = 8\beta\frac{A-2Z-1}{A(A+2)}, \tag{3.16}$$

and, numerically, for the representative values of A and Z already quoted, $\Delta_s = 0{\cdot}238$ MeV. and $\Delta_s = 0{\cdot}202$ MeV., respectively. In total, for the typical pair of neighbouring species for which $A-2Z=11$, $\Delta_e+\Delta_s=0{\cdot}43$ MeV., and for the corresponding pair for which $A-2Z=23$, $\Delta_e+\Delta_s=0{\cdot}37$ MeV. The discrepancy between the purely empirical values and the semi-theoretical values has now been removed—and we may conclude that the process of progressive nuclear synthesis in the sequences

$$^{71}_{30}\text{Zn to } ^{89}_{39}\text{Y } (A-2Z=11) \quad \text{and} \quad ^{123}_{50}\text{Sn to } ^{141}_{59}\text{Pr } (A-2Z=23)$$

is entirely regular (judged against the norm of the stable species throughout the whole range of Z), such structural discontinuities as there may be in these sequences being of trivial importance.

We turn now to the curves for $A-2Z=13$, 15 on fig. 24 and $A-2Z=25$, 27 on fig. 25. These curves represent obvious discontinuities, in the first figure about $N=50$ and in the second about $N=82$. But they also exhibit a doubling, significant of a distinction between disintegrations of the types $\binom{o}{e}\rightarrow\binom{e}{o}$ and $\binom{e}{o}\rightarrow\binom{o}{e}$ not evident for $A-2Z=11$ and 23, which doubling in each case is more pronounced above the discontinuity in N than below. This last effect was noted by Suess (1951) for the two curves which he plotted ($A-2Z=13$, 25). For all four curves it may be said (the points on the curve for $A-2Z=15$ carry larger probable errors than do those on the other curves) that the discontinuity is of smaller magnitude in the branch of the curve belonging to parent species of the type $\binom{o}{e}$ than it is in the other branch $\left(\text{for parent species of the type } \binom{e}{o}\right)$. This feature alone is easy enough to interpret. If we write $B_p\binom{A}{Z}$ for the binding energy of the last-added proton in the nucleus $\binom{A}{Z}$, and similarly $B_n\binom{A}{Z}$ for the binding energy of the last-added neutron in the same nucleus, it is not difficult to show (Feather 1952*a*) that (3.14) may be re-written

$$T_{\bar{\beta}}\binom{A}{Z} - T_{\bar{\beta}}\binom{A+2}{Z+1}$$
$$= B_n\binom{A+3}{Z+2} - B_p\binom{A+3}{Z+2} - \left[B_n\binom{A+1}{Z+1} - B_p\binom{A+1}{Z+1}\right], \quad (3.17)$$

again neglecting as insignificant the difference of the small quantities ϵ'. In the same way we have

$$T_{\bar{\beta}}\binom{A+4}{Z+2} - T_{\bar{\beta}}\binom{A}{Z}$$
$$= B_n\binom{A+1}{Z+1} - B_p\binom{A+1}{Z+1} - \left[B_n\binom{A+5}{Z+3} - B_p\binom{A+5}{Z+3}\right]. \quad (3.18)$$

This last equation gives directly the magnitude of the discontinuity in either branch of each of the four curves which we are now discussing. It will be seen that the relevant energy difference may be regarded as made up of the sum of two terms, one being the difference between the binding energies of two protons (each odd-numbered, or each even-numbered) and the other the difference between the binding energies of two neutrons (each even-numbered if the protons are odd-numbered and vice versa).† Making the reasonable assumption that the features of discontinuity with which we are dealing are neutron-number dependent, we then concentrate attention on the second of these terms to the exclusion of the first. The effect may be seen most clearly in a particular example. We choose for this purpose the comparison of the $\binom{o}{e}$ discontinuity between $^{85}_{36}Kr$ and $^{89}_{38}Sr$ and the $\binom{e}{o}$ discontinuity between $^{87}_{37}Rb$ and $^{91}_{39}Y$ $(A-2Z=13)$. The magnitude of the former discontinuity is $0{\cdot}77 \pm 0{\cdot}02$ MeV., that of the latter $1{\cdot}4 \pm 0{\cdot}1$ MeV. In the former case this involves the contribution $B_n\binom{86}{37} - B_n\binom{90}{39}$, in the latter the contribution $B_n\binom{88}{38} - B_n\binom{92}{40}$, and we conclude that in nuclei for which $A-2Z=12$ the difference between the binding energies of the 49th and 51st neutrons is less than that between the binding

† In the case which we are considering, that is when A is odd.

energies of the 50th and 52nd neutrons by about 0·6 MeV. From fig. 25 we similarly conclude that the difference between the binding energies of the 81st and 83rd neutrons is less than that between the binding energies of the 82nd and 84th neutrons (all in nuclei for which $A-2Z=26$) by about $0{\cdot}45 \pm 0{\cdot}25$ MeV. From fig. 25, also, the indications are that the corresponding difference is somewhat greater in respect of nuclei for which $A-2Z=24$.

We have now interpreted the main feature of discontinuity exhibited by the curves of figs. 24 and 25. Minor irregularities in these curves may still reflect imperfections in experimental knowledge rather than peculiarities of nuclear structure, but attention might be directed to two of them as less likely to be explained in this way than are some of the others. The two points for the β-disintegration energies of the zirconium isotopes $^{93}_{40}Zr$ and $^{95}_{40}Zr$ on fig. 24 are both low, the latter markedly so. This may indicate a Z-dependent effect which we have so far neglected in discussing the data in question.

Fig. 26, as already stated, covers the range of odd-A species having proton numbers in the region $Z \sim 50$. On this figure the doubling of the curves, separating parent species of the parity types $\begin{pmatrix} o \\ e \end{pmatrix}$ and $\begin{pmatrix} e \\ o \end{pmatrix}$ is very pronounced. Furthermore, whilst the curves for parent species of the type $\begin{pmatrix} e \\ o \end{pmatrix}$ show a marked change of slope around $Z=50$, those for parent species of the type $\begin{pmatrix} o \\ e \end{pmatrix}$ do not. We can interpret the fact that the downwards slope of the former curves in this region is greater than that of the latter, just as we have interpreted the discontinuities around $N=50$ and $N=82$ in figs. 24 and 25. We reverse signs in (3.18), and now assume that neutron binding energies are regular within the range of species concerned. We are then left with $B_p\begin{pmatrix} A+1 \\ Z+1 \end{pmatrix} - B_p\begin{pmatrix} A+5 \\ Z+3 \end{pmatrix}$ as the 'irregular' term in the expression for the decrease in T_{β}^{-} with increasing A and Z ($A-2Z$ constant). For $\begin{pmatrix} e \\ o \end{pmatrix}$ parent species this terms involves the difference between the binding energies of the 50th and 52nd protons, for $\begin{pmatrix} o \\ e \end{pmatrix}$ species between the binding energies of the 49th and 51st protons, in the critical range. Empirically, as we see, the former difference is somewhat greater than the latter, as in the similar case of neutron binding.

The fact that the greatest change of slope of the curves of fig. 24 occurs at the plotted point for $N=50$, whereas the greatest change

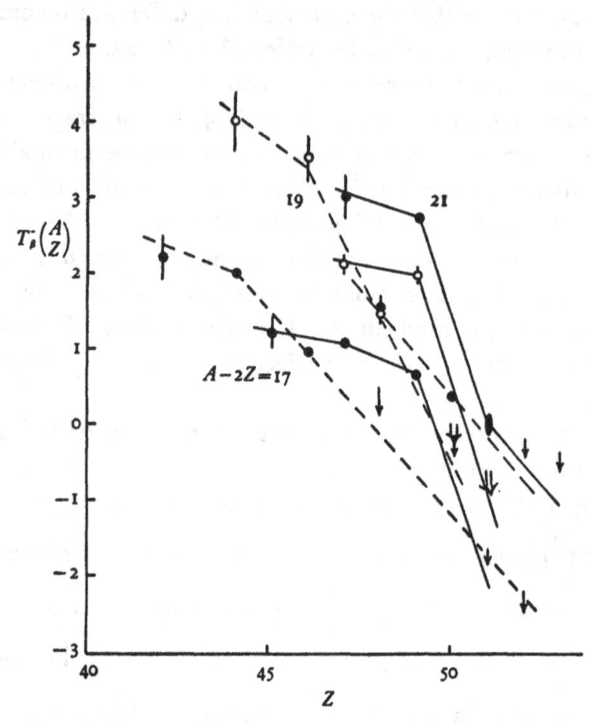

Fig. 26. Energy of β-disintegration as a function of Z, for $A-2Z=17$, 19 and 21.

$T_{\bar{\beta}}\binom{A}{Z}$ in MeV.

of slope in fig. 26 takes place at the point for $Z=49$ finds a natural interpretation when (3.18) is used in this way for the interpretation of both. A little consideration will show that in each case the most abrupt change represents the difference between the decrease in binding energy as between the 48th and 50th nucleon and the decrease as between the 50th and 52nd nucleon of the same type, when $A-2Z$ is constant. In each case, then, the primary discontinuity provides direct evidence for shell closure at the 'magic' number 50—and we must conclude that although completion of this shell takes place in nuclei containing very different numbers of particles (e.g. in $^{87}_{37}Rb$ and $^{119}_{50}Sn$, for neutron and proton shell closure, respectively), neutrons and protons behave very similarly in respect of shell completion.† Though there may be only 37 protons present when the neutron shell of 50 is completed, or as many as 69 neutrons present when the corresponding proton shell is completed, the variation of binding energy with nucleon number in the near-completed shell follows essentially the same pattern in the two cases.

In constructing figs. 24–26 it was possible in each case to isolate a range of species for which, to a good approximation at least, discontinuities in structure involve one type of nuclear constituent only. That is no longer the case when the range of species centred on the proton number $Z=82$ is considered. In this range of species closure of a neutron shell at $N=126$ is believed to occur, on other evidence, and we must expect the situation to be complicated on that account. Fig. 27 presents the experimental information, showing the variation of T_β^- with N for the odd values of $A-2Z$ from 41 to 53 inclusive. It should be pointed out that precise information regarding negative values of T_β^- may be obtained for species within the range with which we are here concerned by applying the method of 'closed disintegration cycles' (see § 2.6) to the detailed results regarding α-disintegration energies which are now available for the heavy radioelements. And negative values may also be deduced by a critical evaluation of possibilities in relation to the capture disintegration of neighbouring isobars of higher Z (see Feather, 1952*b*).

The outstanding feature of fig. 27 is that a marked discontinuity appears only in the curve for $A-2Z=43$, although the 'critical' range of N is also covered by the curve for $A-2Z=45$. For all values of isotopic number except 43 clear doubling of the curves is evident, the branch for parent species of the type $\begin{pmatrix} e \\ o \end{pmatrix}$ lying higher on the figure in each case than the $\begin{pmatrix} o \\ e \end{pmatrix}$ branch.

Paradoxically, the interpretation of the monotonic variation of T_β^- with N for each branch of the curve for $A-2Z=45$ is more clear-cut than is the interpretation of the major discontinuity on the

† See also Coryell, Brightsen and Pappas, 1952.

curve for $A-2Z=43$. Reversing signs in (3.18), as before, we note that the decrease in $T_{\bar{\beta}}$, from point to point along either branch $\binom{o}{e}$ or $\binom{e}{o}$, of a curve for which $A-2Z$ is constant, is given by the difference between the binding energies of the $(Z+1)$th proton and the $(A-Z)$th neutron minus the difference between the binding energies of the $(Z+3)$th proton and the $(A-Z+2)$th neutron in the sequence of nuclei involved. The pairs of neutron and proton numbers concerned in these differences themselves differ by $A-2Z-1$ in each case. For each branch of the curve for $A-2Z=45$ in fig. 27, therefore, binding energies of the Pth proton and $(P+44)$th neutron are involved symmetrically. Now the 'magic' numbers 126 and 82 differ by precisely 44. Thus the (double) curve for $A-2Z=45$ is unique amongst the curves in fig. 27: for the interpretation of this curve, and of this curve alone in the whole figure, we are dealing in imagination with the progressive synthesis of heavier from lighter nuclei under conditions which maintain protons and neutrons 'in step' so far as shell closure is concerned. The smoothness of each branch of this particular curve then shows how closely the last stage of filling the shell of 82 protons, and the first stages of adding extra protons to this structure, match the last stage of filling the shell of 126 neutrons and the subsequent stages of neutron addition thereto, in the series of nuclei for which $A-2Z=44$.

Having reached this conclusion, namely the conclusion that the discontinuities in respect of neutron shell closure entirely balance out those due to proton shell closure on the curve for $A-2Z=45$, we may go further and inquire how the slope of this curve (that is of the two parallel branches of the curve) compares with that predicted on the basis of (3.15) and (3.16). Taking $A=209$, $Z=82$ as representative values, and $r_0=1{\cdot}44\times10^{-13}$ cm., $\beta=20$ MeV., as before, we have $\Delta_e=0{\cdot}148$ MeV., $\Delta_s=0{\cdot}160$ MeV., giving a total slope of 0·31 MeV. per added proton-neutron pair. In fig. 27 the initial slope of the curve for $A-2Z=45$ is 0·48 MeV., the final slope 0·34 MeV. per added nucleon pair. The final slope which has the predicted value, within acceptable limits, here corresponds to the situation when proton and neutron shells are already closed and the first nucleons of the succeeding shells have been added, the initial

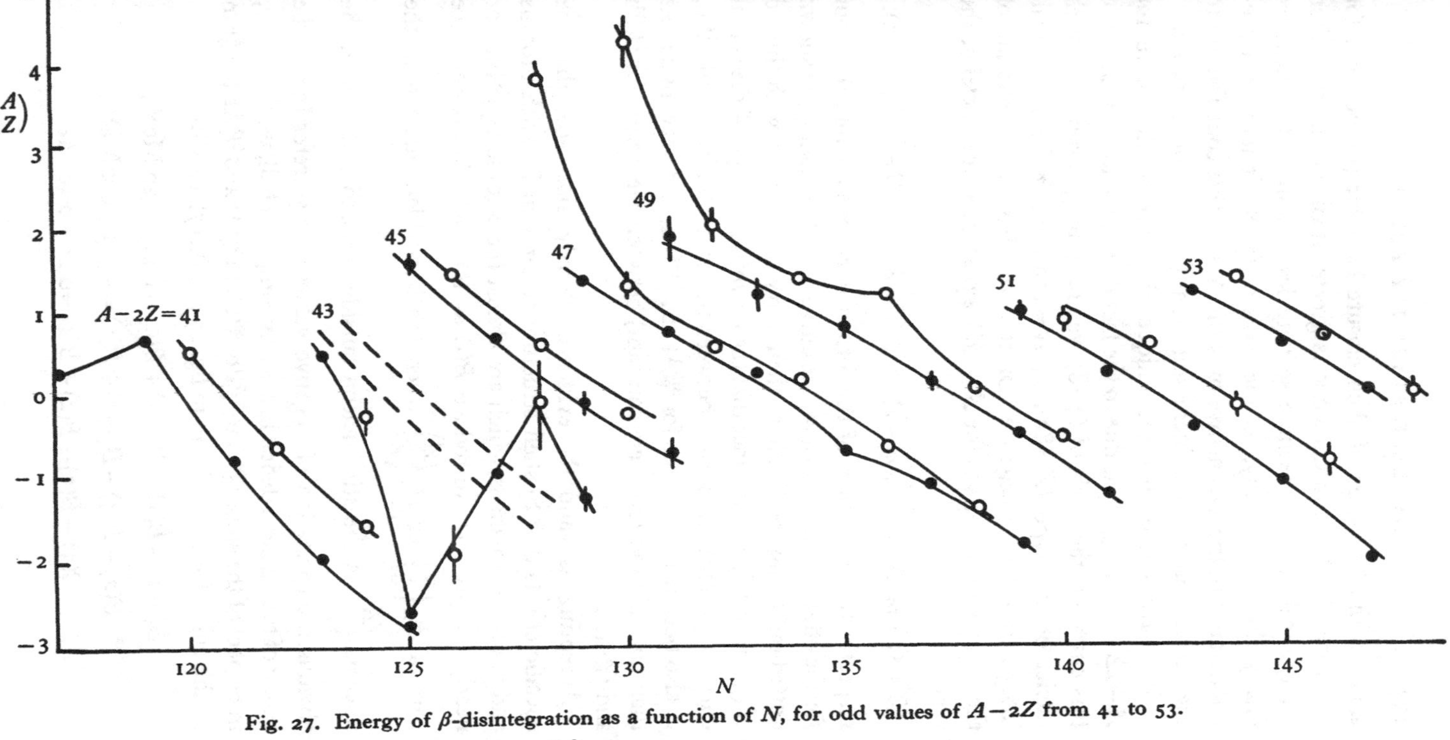

Fig. 27. Energy of β-disintegration as a function of N, for odd values of $A-2Z$ from 41 to 53.

$T_{\beta}^{-}\binom{A}{Z}$ in MeV. ○ odd Z, ● even Z.

slope to the last stage of shell closure both for protons and for neutrons. That the initial slope is greater than normal may then imply a real pause in the increase in nuclear volume as the shells are finally closed: certainly, on the basis of (3.18), it implies a preponderance of the Z-sensitive over the N-sensitive contribution to the decrease in T_{β}^{-} with increasing A, at this stage of the process.

We have stated that the unique character of the curve for $A-2Z=45$ on fig. 27 is due to the fact that in the interpretation of this curve on the basis of (3.18) we are dealing throughout with nuclei of the type ($Z=82+m$, $N=126+m$), m being a small integer, positive or negative. On the same basis interpretation of the neighbouring curves for $A-2Z=43$, 47 involves nuclei of the types

$$(Z=82+m,\ N=126+m-2) \text{ and } (Z=82+m,\ N=126+m+2),$$

respectively. In respect of the former curve we assume that, but for the effects of shell closure, the curve would be double, as shown dotted on the figure. We have then to try to interpret, with the aid of (3.17) and (3.18), the fact that the downwards slope of the actual curve is greater than normal between the points for $^{205}_{81}$Tl and $^{207}_{82}$Pb, as also between the points for $^{213}_{85}$At and $^{215}_{86}$Em, and very much less than normal (and relatively constant) between the points for $^{207}_{82}$Pb and $^{213}_{85}$At.

Accepting as valid the evidence, already discussed, that the specific effects due to shell structure obtain equally with protons and neutrons, in shells which are almost (or completely) 'full' or nearly 'empty', we introduce $B(s)$ as the shell-closure-sensitive part of $B_n\binom{A}{Z}$ or $B_p\binom{A}{Z}$ when the last-added nucleon in the nucleus $\binom{A}{Z}$ is the sth nucleon outside a closed shell. With the obvious interpretation of negative values of s as referring to the extent of the nucleon deficit in a nearly complete shell, we then have, in respect of the steps $^{205}_{81}$Tl to $^{207}_{82}$Pb, $^{207}_{82}$Pb to $^{213}_{85}$At and $^{213}_{85}$At to $^{215}_{86}$Em on the curve for $A-2Z=43$ in fig. 27 (cf. (3.17), (3.18)):

$$B(-1)-B(1)-B(-2)+B(0)=1{\cdot}46\pm0{\cdot}20 \text{ MeV.},$$
$$B(2)-B(4)-B(-1)+B(1)=-3{\cdot}3\pm0{\cdot}5 \text{ MeV.},$$
$$B(3)-B(5)-B(2)+B(4)=0{\cdot}25\pm0{\cdot}50 \text{ MeV.}$$

Writing q for $B(-2)-B(0)$, we obtain:

$$\begin{aligned} B(-1)-B(1) &= q+1{\cdot}46\pm0{\cdot}20 \text{ MeV.,} \\ B(2)-B(4) &= q-1{\cdot}8\pm0{\cdot}55 \text{ MeV.,} \\ B(3)-B(5) &= q-1{\cdot}55\pm0{\cdot}55 \text{ MeV.;} \end{aligned} \qquad (3{\cdot}19)\dagger$$

and we note in particular that $B(5)>B(3)$ unless $q>1{\cdot}55\pm0{\cdot}55$ MeV. On general grounds it is very unlikely that $q>1{\cdot}5$ MeV., thus we conclude that additional nucleons up to at least the fifth in a new shell are less than normally tightly bound (see also Wapstra, 1952). Moreover, if $q>0$ (as it would appear to be if the general credibility of the results in (3·19) is taken as guide), then it also appears that the last-bound nucleon in a completed shell is not necessarily the most tightly bound.

Consistent with the above conclusions are those which may be drawn by applying the same analysis to the first two points on each of the branches of the curve for $A-2Z=47$. For the upper branch of this curve we have qualitatively

$$B(4)-B(2)-B(2)+B(0)>0,$$

and for the lower branch

$$B(5)-B(3)-B(3)+B(1)\sim0.$$

Both these results are easily understandable. So far as the first is concerned, almost certainly $B(0)>B(2)$—and we have already concluded that $B(4)>B(2)$; and the second conforms with our previous general conclusion that $B(5)>B(3)>B(1)$. In complete analogy, the first two points of each of the branches of the curve for $A-2Z=49$ give

$$B(6)-B(2)-B(4)+B(0)>0,$$

and

$$B(7)-B(3)-B(5)+B(1)\sim0;$$

results which are equally understandable on the same basis.

With these remarks made, only one obvious feature of discontinuity in fig. 27 remains unalluded to. This is the step in the lower branch of the curve for $A-2Z=41$ between the points for $^{193}_{76}Os$ and $^{197}_{78}Pt$. Assuming the experimental results to be reliable, the appearance of this discontinuity strongly suggests the conclusion

† Probable errors are not independent.

that the binding energy of the 119th neutron is considerably less than that of the 117th neutron, when $A-2Z=40$. On this basis 118 takes on the status of one of the minor 'magic' numbers.

Finally, we return to the question of the doubling of the curves for $A-2Z$ constant on figs. 24, 25 and 27, to which frequent reference has already been made in a rather unsystematic way. We note that in the range extending for some distance above a discontinuity, at any rate, each branch referring to disintegrations of the type $\binom{e}{o} \to \binom{o}{e}$ lies above the corresponding branch for disintegrations of the type $\binom{o}{e} \to \binom{e}{o}$, the mean separation of the two branches of any curve being of the order of 0·4 MeV. When this applies we have

$$T_{\bar{\beta}}\binom{A}{Z} - T_{\bar{\beta}}\binom{A+2}{Z+1} > T_{\bar{\beta}}\binom{A+2}{Z+1} - T_{\bar{\beta}}\binom{A+4}{Z+2},$$

both A and Z being odd, and $A-2Z$ constant. From (3.17), therefore,

$$B_p\binom{A+5}{Z+3} + B_p\binom{A+1}{Z+1} - 2B_p\binom{A+3}{Z+2}$$
$$> B_n\binom{A+5}{Z+3} + B_n\binom{A+1}{Z+1} - 2B_n\binom{A+3}{Z+2}. \quad (3.20)$$

In (3.20), all mass and charge numbers being even except $Z+2$, the nuclei $\binom{A+1}{Z+1}$ and $\binom{A+5}{Z+3}$ referred to in the inequality are $\binom{e}{e}$ nuclei, and the nucleus $\binom{A+3}{Z+2}$ is an $\binom{o}{o}$ nucleus. The inequality then implies that the binding energy of an odd-numbered proton, in a nucleus of even isotopic number, is less than the mean of the binding energies of the neighbouring even-numbered protons, in nuclei of the same isotopic number, by an amount which is greater than the corresponding deficit characteristic of the last-bound neutrons in the same nuclei. In less precise phraseology, the odd-even distinction is somewhat greater (as represented by a binding energy difference of about 0·4 MeV.) for protons than it is for neutrons in medium-heavy and in heavy nuclei (see also Kohman, 1952).

When $A-2Z$ is even, as already noted (p. 122), the doubling of the curves of T_{β}^{-} against A (or N, or Z) is much more pronounced than when $A-2Z$ is odd. From fig. 23 we observe that the separation of the two branches of the curve for $A-2Z=0$ is about 13·5 MeV., that of the two branches for $A-2Z=2$ decreases from 14 to about 7 MeV. as A increases from 8 to 40, and the separation for $A-2Z=4$ is about 6 MeV. For $A-2Z=46$ and $A-2Z=52$ (see Feather (1952*b*) for the plotted curves) the separation of the branches is practically constant at 2·65 MeV. A simple interpretation of this general effect has been given previously (p. 123): it may be re-stated here in terms of (3.20). For this inequality now, A being even and Z odd, the neutron-proton parities of the nuclei $\binom{A+1}{Z+1}$ and $\binom{A+5}{Z+3}$ are $\binom{o}{e}$, and the nucleus $\binom{A+3}{Z+2}$ is an $\binom{e}{o}$ nucleus. In this case, the magnitude of the inequality involves the excess of the total binding energy of four even-numbered nucleons over the total binding energy of four odd-numbered nucleons (two neutrons and two protons in each group), rather than the difference of the sums, for two groups of nuclei, of the small differences between the binding energies of corresponding protons and neutrons, as before. This distinction having been established, there is no doubt that the greater separation of the branches of the curves for even isotopic number is fully explained on the simple assumption (loosely stated) that the binding energy of even-numbered nucleons is in general greater than that of odd-numbered nucleons. For the heaviest nuclei ($Z \geqslant 86, N \geqslant 130$), the numerical values of separations already given lead to the conclusion that, apart from minor irregularities, the binding energy of an even-numbered proton is greater than the mean of the binding energies of neighbouring odd-numbered protons by 1·5 MeV., and the binding energy of an even-numbered neutron greater than the mean of the binding energies of neighbouring odd-numbered neutrons by 1·1 MeV., in sequences for which $A-2Z$ is constant.

The regularities which we have detected in discussing the experimental data exhibited in figs. 23–27 throw considerable light on certain empirical stability rules which can be traced back to what were no more than enlightened guesses in the early days of classical radioactivity. In the form given it by the writer (Feather, 1943), the rule most directly related to our present discussions may be stated 'In the sequence of disintegrations

$$\binom{A}{Z}\rightarrow\binom{A}{Z+1}\rightarrow\binom{A}{Z+2} \quad\text{or}\quad \binom{A}{Z}\rightarrow\binom{A}{Z-1}\rightarrow\binom{A}{Z-2},$$

when Z is even, the first disintegration energy is less than the second if A is even, and the second disintegration energy is less than the first if A is odd.'† It might be added to this (though it was not added in the original statement of the rule—see Feather (1944 *b*)) that the difference between the two disintegration energies is in general considerably less when A is odd than when A is even. For the discussion of this rule (3.17) becomes

$$T_{\beta}^{-}\binom{A}{Z}-T_{\beta}^{-}\binom{A}{Z+1}$$
$$=B_p\binom{A+1}{Z+1}-B_n\binom{A+1}{Z+1}-B_p\binom{A+1}{Z+2}+B_n\binom{A+1}{Z+2}, \quad (3.21)$$

and, for positron emission, we have similarly

$$T_{\beta}^{+}\binom{A}{Z}-T_{\beta}^{+}\binom{A}{Z-1}$$
$$=B_n\binom{A+1}{Z}-B_p\binom{A+1}{Z}-B_n\binom{A+1}{Z-1}+B_p\binom{A+1}{Z-1}. \quad (3.22)$$

Here the first point to note is that, since the right-hand member of (3.22) may be obtained from the right-hand member of (3.21) by reversing the order of the terms and writing $(Z-2)$ for Z, any result in the form of an inequality depending on the neutron-proton parities of the parent nucleus $\binom{A}{Z}$ will be equally true for the successive positron disintegrations, as for the successive negative β-disintegrations of this nucleus. This being established, from (3.21) alone we see that, when both A and Z are even, the difference between the energies of successive disintegrations (β^- or β^+) is given by the sum of the binding energies of an odd-numbered proton

† It is a direct consequence of this rule that the mass number of any unidentified isotope of an element of even Z is odd, if the isotope is β-active but does not give rise to a daughter product of greater energy release. Thus, if the already reported radioactivity of neodymium ($Z=60$) is confirmed, it must be attributed to a Nd isotope of odd A.

and an odd-numbered neutron, minus the sum of the binding energies of the next (even-numbered) proton and the next (even-numbered) neutron. Again, when A is odd and Z even, the difference between the energies of successive disintegrations is given by the difference between the binding energy of an even-numbered neutron and the binding energy of the next (odd-numbered) neutron minus the corresponding difference for an even-numbered and the preceding (odd-numbered) proton. It may be remarked here that the binding energies concerned in these statements are those characteristic of nuclei for which, in each case, A, not $A-2Z$ is constant, as they have been in previous statements in our recent discussion. We cannot then take over earlier results exactly, but without risk of error we can certainly interpret the fact that, when both A and Z are even, $T_\beta^-\binom{A}{Z} - T_\beta^-\binom{A}{Z+1} < 0$ by postulating that both the even-numbered proton and the even-numbered neutron are more tightly bound than the preceding odd-numbered proton and odd-numbered neutron, respectively. Similarly, the empirical result roughly expressed by

$$\left| T_\beta^-\binom{A}{Z} - T_\beta^-\binom{A}{Z+1} \right| \begin{matrix} A \text{ even} \\ Z \text{ even} \end{matrix} > \left| T_\beta^-\binom{A}{Z} - T_\beta^-\binom{A}{Z+1} \right| \begin{matrix} A \text{ odd} \\ Z \text{ even} \end{matrix}$$

may be understood, on the basis of our present analysis, without detailed discussion. Such discussion is required only if we are to understand why, for A odd and Z even, $T_\beta^-\binom{A}{Z} - T_\beta^-\binom{A}{Z+1}$ is generally positive. This result obviously depends upon the fact that for A constant the tendency is for the proton binding energy to decrease with increasing Z, due to the electrostatic interaction, quite apart from its alternation due to the spin-pairing of particles. The alternation in the case of neutron binding, on the other hand, is superimposed on a steady trend much more nearly approaching constancy.

At this point it is interesting and instructive to relate the empirical rules regarding disintegration energies in successive β-disintegration to the earlier-recognized regularities (existence rules) which our discussion of the systematics of stable species has already revealed (cf. Chapter I). To do this it is necessary only to include negative values of disintegration energy in our survey, and to regard the prohibition of the existence of neighbouring stable isobars as absolute. Then, for even A, we conclude that if there are neighbouring isobars, one stable and the other unstable (against negative or positive β-emission), it is the species of even Z which is stable. And, for odd A, our conclusion must be given under two heads. First, if the odd A isobar of lower Z is stable and the neighbouring isobar of higher Z is β^--active (or the isobar of higher Z is stable and the neighbouring isobar of lower Z is β^+-active) then Z for the stable isobar is odd. Secondly, if the odd A isobar of higher Z is stable and the neighbouring isobar of lower Z is β^--active (or the isobar of lower Z is stable and the neighbouring isobar of higher Z is β^+-active) then Z for the stable isobar may be either odd or even. A little consideration will show that these conclusions re-formulate exactly our previous statement that, whilst examples of the $\binom{A}{Z}$ parity type $\binom{\eta}{\omega}$ are not found amongst stable species, examples of each of the other three parity types are of frequent occurrence (Feather, 1946). The only proved exceptions, as already noted, are for $Z \leqslant 7$.

Obviously the empirical rules which we have been discussing lose all basis of 'theoretical' justification when the energy relations for the successive disintegrations are complicated by the incidence of shell closure in the nuclei involved. This aspect of the matter may be followed for the case A odd, Z even by reference to fig. 27. On this figure the representative point for the 'initial' nucleus $\binom{A}{Z}$ in this case lies on the lower branch of the curve for $A-2Z$ constant. Similarly, the representative point for the 'first daughter' nucleus $\binom{A}{Z+1}$ falls on the upper branch of the curve for $A-2Z-2$ and one unit in N to the left of the other, and the point for the 'second daughter' nucleus $\binom{A}{Z+2}$, if the sequence is extended, falls on the lower branch of the curve for $A-2Z-4$ and one unit in N farther to the left, and so on. So far as fig. 27 is concerned there is only one clear instance in which our rule for A odd, Z even is broken: the energy of capture transformation of $^{207}_{84}Po$ is less than the energy of capture transformation of $^{207}_{83}\mathrm{Bi}$ by about 1 MeV. There is a second doubtful instance: for the successive capture transformations $^{213}_{86}\mathrm{Em} \rightarrow {}^{213}_{85}\mathrm{At} \rightarrow {}^{213}_{84}\mathrm{Po}$ it may well be that the energy available for the second transformation is greater than for the first (or the first may be energetically disallowed)—only further experiment will decide.

Leaving aside the very few instances in which the rule for A odd, Z even is broken, we may discuss the extension of the rule—and the interpretation of the rule as so extended—also with the aid of fig. 27. From what has just been stated regarding the relative positions of the representative points for successive members of the isobaric sequence $\binom{A}{Z}, \binom{A}{Z+1}, \binom{A}{Z+2}, \ldots$, and from the lie of the various curves on the figure, we have, for A odd (and Z odd or even)

$$T^{\pm}_{\beta}\binom{A}{Z} > T^{\pm}_{\beta}\binom{A}{Z \mp 1} > T^{\pm}_{\beta}\binom{A}{Z \mp 2} > \ldots \qquad (3.23)$$

(see Sitte, 1948). Given that the curves of T^{-}_{β} against N, for $A-2Z$ constant, are roughly parallel, and are of negative slope throughout (shell-closure effects excluded), a little consideration will show that (3.23) implies that the separation of the lower branch of any such curve and the higher branch of the curve of next lower (odd) $A-2Z$ is greater than one unit parallel to the axis of N. An equivalent statement (in respect of the portion of the figure below the line $T^{-}_{\beta}=0$) is that if the species $\binom{A}{Z}$ of odd isotopic number is stable against negative β-disintegration then the species $\binom{A}{Z+1}$, of isotopic number two units lower than the other, is also β^{-}-stable.

For species of even isotopic number we have already seen that the separation of the two branches of the curve of T^{-}_{β} against N ($A-2Z$ constant) is much larger than the corresponding separation when the isotopic number is odd. Even for the heaviest species (p. 137) this separation is greater than that of the upper (or, alternatively, lower) branches of the curves for neighbouring even values of $A-2Z$ (in this range, in fact, the lower curve for $A-2Z$ (even) coincides roughly with the upper curve for $A-2Z-4$, see Feather (1952*b*)). In this result the general validity of the rule for A even, Z even is demonstrated graphically, and, from it, its extension to cover more than two successive disintegrations follows simply. We obtain, for comparison with (3.23),

$$\left.\begin{aligned} &T^{\pm}_{\beta}\binom{A}{Z} < T^{\pm}_{\beta}\binom{A}{Z\mp 1} > T^{\pm}_{\beta}\binom{A}{Z\mp 2} < T^{\pm}_{\beta}\binom{A}{Z\mp 3} > \dots,\\ &T^{\pm}_{\beta}\binom{A}{Z} > T^{\pm}_{\beta}\binom{A}{Z\mp 2} > \dots,\\ &T^{\pm}_{\beta}\binom{A}{Z\mp 1} > T^{\pm}_{\beta}\binom{A}{Z\mp 3} > \dots. \end{aligned}\right\} \quad (3.24)$$

Inequalities (3.23) and (3.24) find most extensive application to the series of successive β-disintegrations observed with the products of fission. From information at present available it appears that disintegration energies are known for the last two members of 25 series of 'genetically related' fission products of odd A, for the last three members of 4 series of fission products of the same type, and for the last two members of 10 series of genetically related fission products of even A. In the first group of 25 pairs the first product is of even Z in 18 cases and of odd Z in 7 cases; in the last group (even A) the first fission product is of even Z in all cases. Over all there is only one clearly anomalous case: the disintegration energy of $^{117}_{48}Cd$ is less than that of $^{117}_{49}In$. This anomaly is obviously related to the incidence of proton shell closure at $_{50}Sn$: applying (3.21) to the two active species concerned we have

$$B_p\binom{118}{50} - B_p\binom{118}{49} > B_n\binom{118}{50} - B_n\binom{118}{49}.$$

Each member in this inequality is almost certainly positive: the fact that the sign of the inequality is opposite to the normal sign then derives most simply from the assumption that $B_p\binom{118}{50}$ is unusually large. One anomaly, and that to be expected, in 39 cases represents a very complete measure of conformity between fact and empirical rule. Near-anomalies may be noted in three other cases, though the rule is not definitely broken. The disintegration energies of $^{95}_{40}Zr$, $^{97}_{40}Zr$ and $^{141}_{56}Ba$ all appear low, in the light of (3.23), by comparison with the disintegration energies of the respective daughter products. If this appearance is significant (cf. p. 129) it probably indicates unexpectedly low values for $B_p\binom{96}{41}$, $B_p\binom{98}{41}$ and $B_p\binom{142}{57}$—as if 40 and 56 were minor 'magic' numbers in respect of Z.

In any survey of published information regarding fission product 'chains', it will be evident at once that information is much less extensive as regards chains of even A than it is for the genetically related series of products of odd mass number.† This result is in part to be attributed to the fact (see fig. 1) that in any small range of A the stable species of greatest neutron excess are of even A and Z. But another fact enters also. Since the stable end-product of a chain for which A is even is of even Z, the last member of the chain, according to (3.24), is likely to have greater disintegration energy, on the average, than the last member of a chain of odd A (see (3.23)). 'Chains' involving a single member are, statistically, more frequent for even A than for odd A, on the basis of fig. 1; when such 'single-member chains' occur they are more likely to escape investigation because of the high energy, and consequent short lifetime, of the active species concerned.

One 'single-member chain' of some interest is the fission product $^{136}_{53}I$ (Stanley and Katcoff, 1949). The maximum β-particle energy in this case is 6·5 MeV. and a γ-ray of 2·9 MeV. has been reported: the disintegration energy is therefore at least 6·5 MeV., and it may be 9·4 MeV.; in either case a very high value for a medium-heavy element. The binding energies involved are $B_p\binom{137}{54}$ and $B_n\binom{137}{54}$. Obviously the binding energy of the 83rd neutron in $^{137}_{54}Xe$ must be very small, on the basis of either assumption. In this connexion it is significant to note that $^{136}_{54}Xe$ is by far the lightest stable nucleus having a neutron excess (isotopic number) of 28: the next lightest such nucleus is $^{148}_{60}Nd$ (see §§ 1.5 and 1.6). But further consideration of the properties of $^{137}_{54}Xe$ belong properly to the next section.

3.4. 'Delayed' neutron emission

The lifetime of a nucleus which is unstable in respect of neutron emission is necessarily extremely short: there is no 'potential barrier' which the outgoing particle must 'penetrate'. Accepting this point of view, the only reasonable explanation of the delayed

† No information has been published concerning fission products having $A = 74$, 76, 80, 96, 98, 100, 104, 110, 116, 120, 122, 124, 128, 130, 148, 150, 152 and 154.

emission of neutrons in fission (Roberts, Meyer and Wang, 1939) is that these neutrons are emitted from excited states of nucl which are not themselves primary products of fission: the 'delayed neutrons are emitted 'instantaneously' (possibly in competitio with γ-rays) from daughter nuclei left excited after β-emission This was the explanation put forward by Bohr and Wheeler (1939 very soon after the phenomenon was discovered. On this basis th experimentally determined half-value periods for delayed neutro decay are the half-value periods characteristic of the β-disintegra tions in which the neutron-unstable states are excited. Delaye neutrons of five different periods all less than 1 min. are we established,†, and three longer periods (about 3, 12 and 120 min respectively) have recently been reported (Kunstadter, Floyd Borst and Weremchuk, 1951). The β-active parents of the delayed neutron emitters of 55·6 and 22·6 sec. effective half-value perio have been identified as $^{87}_{35}$Br and $^{137}_{53}$I, respectively (Snell, Levinger Meiners, Sampson and Wilkinson, 1947), and there is stron presumption that $^{89}_{35}$Br is the parent of the delayed-neutron emitte of effective period 4·5 sec. (Sugarman, 1947). It has been shown that only about 2 % of the β-disintegrations of $^{87}_{35}$Br lead to neutron emission from $^{87}_{36}$Kr* and only about 7 % of the β-disintegrations o $^{137}_{53}$I to neutron emission from $^{137}_{54}$Xe* (Levinger, Meiners, Sampson Snell and Wilkinson, 1950), and the mean energies of the neutron emitted by the two daughter products have been determined a 0·25 and 0·56 MeV., respectively. A detailed investigation of th β- and γ-radiations of $^{87}_{35}$Br (Stehney, 1950) has shown that the tota disintegration energy is 8·0 MeV., but that about 70 % of th disintegrations lead to a state or states in $^{87}_{36}$Kr of 5·4 MeV. excitation energy. From this it is clear that the neutron-emitting state $^{87}_{36}$Kr* has at least this energy of excitation, and, taking the neutron energy measurements into consideration, we conclude that $B_n\binom{87}{36}$ is a least 5·1 MeV.

In the last paragraph we have summarized, very briefly, th present state of our empirical knowledge concerning the emission

† Snell, Nedzel, Ibser, Levinger, Wilkinson and Sampson, 1947; Redman and Saxon, 1947; Hughes, Dabbs, Cahn and Hall, 1948; Creveling, Hood and Pool, 1949; Sun, Charpie, Pecjak and Jennings, 1950; Brolley, Cooper, Hall Livingston and Schlacks, 1951.

of delayed neutrons in fission. We have now to attempt an interpretation of the facts using current theoretical views as guide. In 1944 the writer made two predictions, assuming no more than very general results regarding the dependence of total energy of binding on the neutron-proton parity type of the nucleus concerned (Feather, 1944*b*). These predictions were that delayed neutrons would be found to be more numerous, in relation to 'prompt' neutrons, when the compound nucleus undergoing fission is of odd Z than when it is of even Z, other factors being constant, and, secondly, that the 'delayed' neutrons were more likely to originate in excited fragment nuclei of even Z than of odd Z.† Concerning the first prediction, published results have at present very little to say, but the second is obviously borne out in the case of the delayed neutron emitters so far identified, namely $^{87}_{36}Kr$, $^{89}_{36}Kr$ and $^{137}_{54}Xe$. A more pertinent consideration was advanced by Mayer (1948), after the first identifications had been made. It was that the excitation necessary for neutron emission is likely to be attained most easily (and thus most often) in daughter nuclei containing just one neutron outside a closed shell—and in this group of nuclei most easily, when the neutron excess is greatest (Z least). Both conditions are satisfied with $^{137}_{54}Xe$: the neutron number is 83, and we have already noted (p. 143) the anomalous placing of the product nucleus $^{136}_{54}Xe$ in the series of stable nuclei for which $A-2Z=28$. $^{136}_{54}Xe$ is the lightest stable nucleus for which $N=82$. Again, with $^{87}_{36}Kr$, the neutron number is 51 and $^{86}_{36}Kr$ is the lightest stable nucleus having $N=50$. A slightly different situation obtains with $^{89}_{36}Kr$. In this case the neutron number is 53—that is the emitted neutron is the second odd-numbered neutron outside the closed shell of 50—but the product $^{88}_{36}Kr$ is itself β-active and the lightest stable species having $N=52$ is $^{92}_{40}Zr$.

Having regard to the success of Mayer's interpretation, its use for further predictions is probably justified. There is some evidence that the mass number of the fission product responsible for the delayed neutron activity of 1·5 sec. half-value period lies in the range 129–135 (Sugarman, 1947). If this conclusion is correct, then we may fairly confidently expect that the disintegration sequence

† Later Way and Wigner (1948) concluded independently, though less successfully, that the neutron-emitting nucleus should be of even A and even Z.

involved is ${}^{135}_{51}\mathrm{Sb} \xrightarrow{\beta} {}^{135}_{52}\mathrm{Te}^* \xrightarrow{n} {}^{134}_{52}\mathrm{Te} \xrightarrow{\beta} {}^{134}_{53}\mathrm{I} \xrightarrow{\beta} {}^{134}_{54}\mathrm{Xe}$. Similarly, the sequence ${}^{85}_{33}\mathrm{As} \xrightarrow{\beta} {}^{85}_{34}\mathrm{Se}^* \xrightarrow{n} {}^{84}_{34}\mathrm{Se} \xrightarrow{\beta} {}^{84}_{35}\mathrm{Br} \xrightarrow{\beta} {}^{84}_{36}\mathrm{Kr}$ merits serious consideration in respect of the shortest period definitely established that given as 0·43 sec. in recent reports. So far as the three periods longer than 1 min. are concerned, it is possible that certain of the minor 'magic' numbers enter into their interpretation on the basis of similar considerations. Here we shall permit ourselves only one further speculation. The nucleus ${}^{139}_{53}\mathrm{I}$ occupies exactly the same position in the system of nuclei in relation to the neutron number 82 as ${}^{89}_{35}\mathrm{Br}$ does in relation to $N=50$. It will be rather surprising, therefore, if a component of delayed-neutron radiation cannot be identified having the half-value period of this body ('delayed' neutron emitted in the transition ${}^{139}_{54}\mathrm{Xe}^* \rightarrow {}^{138}_{54}\mathrm{Xe}$). The fact that the period of ${}^{139}_{53}\mathrm{I}$ (2·7 sec.) lies between two already accepted delayed-neutron periods in the 1–5 sec. range may account for the lack of such identification at the present time.

Outside the range of the fission products, two other delayed-neutron emitters are known. In 1949 Alvarez established the fact that neutrons are emitted following the β-disintegration of ${}^{17}_{7}\mathrm{N}$ (half-value period 4·2 sec.), and, more recently, a similar result has been reported for ${}^{9}_{3}\mathrm{Li}$ (half-value period 0·168 sec.) (Gardner, Knable and Moyer, 1951). In the former case Mayer's regularity still holds—the emitted neutron is the single neutron outside the closed shell of 8 neutrons in ${}^{17}_{8}\mathrm{O}^*$; in the latter instance, however, the nucleon number is so small that the concept of shells scarcely applies, but the emitted neutron is the 'last' neutron in ${}^{9}_{4}\mathrm{Be}^*$—and we know that the binding energy of this neutron in the ground state of the nucleus is unusually small ($\sim$ 1·6 MeV.).

In this section we have discussed the facts concerning the emission of neutrons from nuclei excited as the result of previous β-disintegration for the light which they throw on our general notions of the structure of the nucleus. It was first pointed out by the writer (Feather, 1942) that a nucleus which is able to emit in succession a β-particle and a neutron, is also able, so far as energy considerations are concerned, to effect the same overall transformation by the emission of a negative proton, if the mass of this hypothetical particle is less than the sum of the masses of the electron and the neutron (in particular, if its mass is equal to the mass of the positive proton). Experiments to test the possibility that negative protons are emitted from some of the products of fission have been carried out by Broda, Feather and Wilkinson (1947) and by Scott and Titterton (1950). In these experiments it was shown that the branching ratio as between negative-proton- and β-emission, in respect of those product nuclei energetically capable of emitting negative protons, is not greater than 2×10^{-5} (this figure differs from the published figures because more precise information is now available concerning the degree of excitation of the 'delayed'-neutron-emitting states). To this degree of precision, the conclusion is, therefore, that negative protons are not emitted in these 'low-energy' transformations. Michel (1950) has pointed out that any other conclusion would have had very grave consequences for the current (meson) theory of the interaction of nuclear particles.

CHAPTER IV

SPONTANEOUS FISSION AND THE NUMBER OF THE ELEMENTS

The discovery of neutron-induced fission in 1939 drew attention to the fact, at that time widely disregarded, that the heaviest nuclei are energetically unstable in relation to spontaneous division into two fragments of roughly equal mass. Even on the basis of mass determinations then available it was clear, for example, that the nucleus $^{238}_{92}U$ might divide into two (β-unstable) nuclei $^{119}_{46}Pd$ with the release of energy. The mass of the neutral atom $^{238}_{92}U$ was then quoted as 238·14, that of the neutral atom $^{110}_{46}Pd$ as 109·942, and the mass of the neutron as 1·0089. On this basis the division of $^{238}_{92}U$ into two stable $^{110}_{46}Pd$ nuclei and 18 neutrons appeared energetically allowed to the extent of roughly 100 MeV.; its division into two $^{119}_{46}Pd$ nuclei was therefore allowed within an even wider margin of energy. Further, and to a rough approximation, such symmetrical (spontaneous) fission appeared possible for all nuclei having $A > 90$ (or thereabouts)—that is for every species for which the packing fraction of Aston (see p. 77) is greater than the packing fraction of the species of half the mass number concerned. From the practical point of view it became a question for experiment to decide whether the phenomenon was observable, and if so to determine disintegration constants and other quantitative features of the process in particular cases.

The first published account of the detection of spontaneous fission is that of Petrzhak and Flerov (1940). Since these authors gave a provisional value for the effective spontaneous fission disintegration constant of natural uranium, a great deal of work has been done both with natural products and with separated isotopes. A summary of much of the work, up to the end of 1945, is given in a review by Segrè (1951). Some results have also been published independently since the end of the war. Here we discuss briefly certain general features of the phenomenon by way of introduction, and then consider in slightly more detail the information at present available concerning disintegration constants and energies.

The neutrons emitted in spontaneous fission were probably first detected by Rotblat early in 1941 (see Rotblat, 1943). Natural uranium was used. Somewhat later Fermi and colleagues detected the same effect, and subsequent investigations have merely confirmed the early results. It can now be stated that the rate of emission of secondary neutrons in the spontaneous fission of uranium is $2{\cdot}2 \pm 0{\cdot}3$ per fission event. This is to be compared with the value $2{\cdot}5 \pm 0{\cdot}1$ secondary neutrons per thermal-neutron induced fission of ${}^{235}_{92}U$ recently released by the declassification authorities of U.S., Great Britain and Canada (see *Canad. J. Phys.* 29, 203 (1951)).

The energy spectrum of the fission fragments produced in the spontaneous fission of natural uranium has been compared with that characteristic of the fragments produced in the thermal-neutron induced fission of ${}^{235}_{92}U$ by Whitehouse and Galbraith (1950). These authors find no significant difference between the two spectra. If there is any such difference it is in the sense that there is a slight lack of fragments of the highest energy (leading to a consequent narrowing of the higher-energy peak) in the spectrum belonging to the fragment nuclei of spontaneous fission.

The detection of specific fission products has not been systematically pursued, but Macnamara and Thode (1950) have identified xenon isotopes of mass numbers 129, 131, 132, 134 and 136, and krypton isotopes of mass numbers 83, 84 and 86, present in pitchblende from the Great Bear Lake region of Canada, as due to the spontaneous fission of uranium. It was concluded that 80% of the xenon in this mineral was of that origin.

Information regarding spontaneous fission disintegration constants is collected in Table XII. In this table, for purpose of comparison, entries referring to species characterized by the same value of $A-2Z$ are set in juxtaposition. Taken as a whole, the results confound prediction in almost every particular (see Turner, 1945). So far as can be seen there is no systematic increase of spontaneous fission probability with decrease of A amongst the isotopes of a given element (see Bohr and Wheeler, 1939), even when species of the same parity type only are considered, and there is hardly any significant, and certainly no regular, increase of λ_f with increasing Z in the sequence of 'representative' species

(p. 48) $^{231}_{91}$Pa, $^{234}_{92}$U, $^{237}_{93}$Np and $^{239}_{94}$Pu. Almost any theory is bound to predict that the spontaneous fission probability increases rapidly as the excitation energy necessary to produce 'prompt' fission of the nucleus decreases, but the results of Koch, McElhinney and Gasteiger (1950) on the photo-fission thresholds of $^{232}_{90}$Th, $^{233}_{92}$U, $^{235}_{92}$U, $^{238}_{92}$U and $^{239}_{94}$Pu are not entirely what one would expect on the basis of this assumption. It is true that these authors find very little difference between the threshold energies for fission of these five species (much less difference than was originally predicted by the theorists), but the differences which they report are such as to lead one to expect that $^{238}_{92}$U would be the most active amongst the five species in respect of this mode of disintegration, whereas in fact $^{239}_{94}$Pu is the most active. Possibly the situation will be further clarified when the threshold energies of $^{238}_{94}$Pu and $^{242}_{96}$Cm have been determined experimentally; for the present, at least, it remains obscure.

TABLE XII

Species	$^{231}_{91}$Pa	$^{233}_{92}$U	$^{230}_{90}$Io	$^{234}_{92}$U
$A-2Z$	49	49	50	50
λ_f (sec.$^{-1}$)	2×10^{-24}	$<10^{-25}$	$<2\times10^{-25}$	$<4\times10^{-24}$
Species	$^{238}_{94}$Pu	$^{242}_{96}$Cm	$^{235}_{92}$U	$^{237}_{93}$Np
$A-2Z$	50	50	51	51
λ_f (sec.$^{-1}$)	$8{\cdot}4\times10^{-19}$	$3{\cdot}1\times10^{-15}$	$(1{\cdot}1\pm0{\cdot}7)\times10^{-25}$	$\ngtr 6\times10^{-25}$
Species	$^{239}_{94}$Pu	$^{232}_{90}$Th	$^{238}_{92}$U	
$A-2Z$	51	52	54	
λ_f (sec.$^{-1}$)	4×10^{-24}	$1{\cdot}6\times10^{-26}$	$2{\cdot}7\times10^{-24}$	

Empirically, there is just one tendency which Table XII appears to reveal. It is that within groups for which $A-2Z$ is constant the variation of λ_f with Z is more pronounced when A (or $A-2Z$) is even than when A is odd (see Seaborg, 1952; Whitehouse and Galbraith, 1952). Amongst the four species of isotopic number 50 represented in the table it seems likely that the spontaneous fission probability increases monotonically with Z—and the experimental values indicate that this increase is by a factor of more than 10^{10} in respect of an increase of 6 units in Z, from $Z=90$ to $Z=96$. Unfortunately, there is not at present comparable information for the sequences $A-2Z=52$ and $A-2Z=54$. If a similar variation were to apply to the former of these sequences we might guess that the spontaneous fission rate of $^{244}_{96}Cm$ is within a factor of 10^{-6} of the α-particle disintegration rate of this body (half-value period 10 years), and if the monotonic increase in the sequence $A-2Z=50$ were to be continued to $^{246}_{98}Cf$ ($\tau \sim 35$ hr.) it might well be that a spontaneous fission rate as great as 10^{-5} of the α-particle rate is to be found with this relatively short-lived species. Increase at the same rate over a further 6 units in Z (to $Z=104$) would reduce the spontaneous fission lifetime to 1 sec. or less, and whatever might then be the properties of the nucleus in respect of α-disintegration it would obviously be on the verge of non-existence in respect of fission. Admittedly this involves rather wild extrapolation, but it leads one to hazard the guess that for species of even A the number of the elements is less than 110. From the meagre information of Table XII we might add to this guess the tentative prediction that for odd A the natural series extends somewhat farther, to higher Z.

REFERENCES AND AUTHOR INDEX

AHRENS (1948). *Nature, Lond.*, **162**, 413. *page*

ALBURGER and FRIEDLANDER (1951). *Phys. Rev.* **81**, 523. 35

ALDRICH and NIER (1948). *Phys. Rev.* **74**, 876. 14

ALLARD (1948). *J. Physique*, **9**, 225. 9

ALLEN, K. W. and DEWAN (1950). *Phys. Rev.* **80**, 181. 76

ALVAREZ (1949). *Phys. Rev.* **75**, 1127. 146

AMBROSINO and PIATIER (1951). *C.R. Acad. Sci., Paris*, **232**, 400. 50

ARROE (1950). *Phys. Rev.* **79**, 836. 80

ASARO, REYNOLDS, F. L. and PERLMAN (1951). American Report, UCRL 1533. 60

AVIGNON (1950). *J. Physique*, **11**, 521. 86

BALLOU (1948). *Phys. Rev.* **73**, 630. 4, 17

BALLOU (1949). *Phys. Rev.* **75**, 1105. 75

BARTLETT (1932). *Nature, Lond.*, **130**, 165. 26

BARTON, GHIORSO and PERLMAN (1951). *Phys. Rev.* **82**, 13. 51

BELL and CASSIDY (1950). *Phys. Rev.* **77**, 409. 4

BÉNÉ, DENIS and EXTERMANN (1950). *Helv. phys. Acta*, **23**, 493. 81

BERTHELOT (1942*a*). *J. Physique*, **3**, 17. 34

BERTHELOT (1942*b*). *J. Physique*, **3**, 52. 34

BERTHELOT (1948). *C.R. Acad. Sci., Paris*, **227**, 829. 34

BESTENREINER and BRODA (1949*a*). *Nature, Lond.*, **164**, 658. 3

BESTENREINER and BRODA (1949*b*). *Nature, Lond.*, **164**, 919. 3

BETHE (1937). *Rev. Mod. Phys.* **9**, 69 (p. 162). 28

BISWAS (1949*a*). *Phys. Rev.* **75**, 530. 34

BISWAS (1949*b*). *Indian J. Phys.* **23**, 51. 34

BLEULER and GABRIEL (1947). *Helv. phys. Acta*, **20**, 67. 4

BOHR (1913). *Nature, Lond.*, **92**, 231. 78

BOHR and WHEELER (1939). *Phys. Rev.* **56**, 426. 144, 149

BOTHE and BECKER (1930). *Z. Phys.* **66**, 307. 35

BRADT and SCHERRER (1945). *Helv. phys. Acta*, **18**, 405. 84

BRAID (1951). Ph.D. thesis. Edinburgh. 103

BREIT (1932). *Phys. Rev.* **42**, 348. 78

BREIT (1934). *Phys. Rev.* **46**, 319. 78

BREIT, ARFKEN and CLENDININ (1950). *Phys. Rev.* **78**, 390. 78, 79

BRIX and FRANK (1950). *Z. Phys.* **127**, 289. 79

BRIX and KOPFERMANN (1948). *Naturwissenschaften*, **35**, 161. 78

BRODA (1947). *J. Sci. Instrum.* **24**, 136. 3

BRODA and FEATHER (1947). *Proc. Roy. Soc.* A, **190**, 20. 45

BRODA, FEATHER and WILKINSON, D. H. (1947). *Report Cambridge Conference*, **1**, 114. 147

BROLLEY, COOPER, HALL, W. S., LIVINGSTON and SCHLACKS (1951). *Phys. Rev.* **83**, 990. 144

BUNYAN, LUNDBY and WALKER (1949). *Proc. Phys. Soc.* A, **62**, 253. 33

BUTEMENT (1951). *Nature, Lond.*, **167**, 400. 17

BUTT and BRODIE (1951). *Proc. Phys. Soc.* A, **64**, 791. 103

CAMPBELL, C. G., HENDERSON, W. J. and KYLES (1952). *Phil. Mag.* **43**, 126. *page* 102
CAMPBELL, N. R. (1923). *The Structure of the Atom*, p. 53. Cambridge University Press. 28
CARLSON (1951). *Phys. Rev.* **83**, 203. 14
CARSS, GUM and POOL (1950). *Phys. Rev.* **80**, 1028. 24
CASSELS, DAINTY, FEATHER and GREEN (1947). *Proc. Roy. Soc.* A, **191**, 428. 76
CASSIDY (1951). *Phys. Rev.* **83**, 483. 24
CECCARELLI, QUARENI and ROSTAGNI (1950). *Phys. Rev.* **80**, 909. 4
CLARK, SPENCER PALMER and WOODWARD (1944*a*). British Report, Br 431. 34
CLARK, SPENCER PALMER and WOODWARD (1944*b*). British Report, Br 522. 34
CLARK, SPENCER PALMER and WOODWARD (1945). British Report, Br 584. 34
COLGATE (1951). *Phys. Rev.* **81**, 1063. 4
COOK, MCMILLAN, PETERSON and SEWELL (1949). *Phys. Rev.* **75**, 7. 82
CORYELL, BRIGHTSEN and PAPPAS (1952). *Phys. Rev.* **85**, 732. 131
CRANSHAW and HARVEY, J. A. (1948). *Canad. J. Res.* A, **26**, 243. 34
CRAWFORD and SCHAWLOW (1949). *Phys. Rev.* **76**, 1310. 78
CREVELING, HOOD and POOL (1949). *Phys. Rev.* **76**, 946. 144
CUER and LATTES (1946). *Nature, Lond.*, **158**, 197. 2
CURRAN, ANGUS and COCKROFT (1949). *Phys. Rev.* **76**, 853. 7
DANCOFF (1950). American Report, AECD 2853. 53
DAUDEL and JEAN (1949). *C.R. Acad. Sci., Paris*, **228**, 662. 10
DAUDEL, JEAN and LECOIN (1947). *J. Physique*, **8**, 238. 8
DE BENEDETTI and KERNER (1947). *Phys. Rev.* **71**, 122. 35
DE BENEDETTI and MINTON (1952). *Phys. Rev.* **85**, 944. 35
DE JUREN (1950). American Report, AECD 2854. 82
DE JUREN and KNABLE (1950). *Phys. Rev.* **77**, 606. 83
DEUTSCH and ROTBLAT (1951). American Report, AECD 3179. 76
DEVONS (1949). *Excited States of Nuclei*, p. 32. Cambridge University Press. 53
DUCKWORTH, BLACK and WOODCOCK (1949). *Phys. Rev.* **75**, 1616. 15, 90
DUCKWORTH, KEGLEY, OLSON and STANFORD (1951). *Phys. Rev.* **83**, 1114. 77
DUCKWORTH, PRESTON, R. S. and WOODCOCK (1950). *Phys. Rev.* **79**, 188. 77
DUCKWORTH, WOODCOCK and PRESTON, R. S. (1950). *Phys. Rev.* **78**, 479. 77
DZHELEPOV (1949). *Doklady Akad. Nauk, S.S.S.R.* **65**, 149. 112
ELLIS and MOTT (1933). *Proc. Roy. Soc.* A, **141**, 502. 92
ELSASSER (1933). *J. Physique*, **4**, 549. 26
ELSASSER (1934*a*). *J. Physique*, **5**, 389. 26
ELSASSER (1934*b*). *J. Physique*, **5**, 635. 26
EVANS and HENDERSON, M. C. (1933). *Phys. Rev.* **44**, 59. 3
FAJANS (1912). *Phys. Z.* **13**, 699. 83
FAJANS (1926). *Naturwissenschaften*, **14**, 963. 40

FARRAGI and BERTHELOT (1951). *C.R. Acad. Sci., Paris*, **232**, 2093. pages 2, 69

FAY, GLÜCKAUF and PANETH (1938). *Proc. Roy. Soc.* A, **165**, 238. 74

FEATHER (1942). Unpublished notes dated 26 January. 146

FEATHER (1943). British Report, Br 312. 137

FEATHER (1944*a*). British Report, Br 499. 34

FEATHER (1944*b*). British Report, Br 503. 138, 145

FEATHER (1945). British Report, Br 640. 4, 34, 84

FEATHER (1946). *Proc. Roy. Soc. Edinb.* A, **62**, 211. 91, 140

FEATHER (1948*a*). *Rep. Progr. Phys.* **11**, 19. 10, 34, 86

FEATHER (1948*b*). *Nature, Lond.*, **161**, 431. 113

FEATHER (1951). *Phil. Mag.* **42**, 568. 46

FEATHER (1952*a*). *Phil. Mag.* **43**, 133. 128

FEATHER (1952*b*). *Proc. Roy. Soc. Edinb.* A, **63**, 242. 69, 114, 131, 137, 141

FEATHER (1952*c*). *Phil. Mag.* **43**, 476. 62

FEATHER and BRETSCHER (1938). *Proc. Roy. Soc.* A, **165**, 530. 84, 94

FEATHER and RICHARDSON (1948). *Proc. Phys. Soc.* **61**, 452. 99, 100, 102

FEENBERG (1947). *Rev. Mod. Phys.* **19**, 239. 126

FEENBERG and TRIGG (1950). *Rev. Mod. Phys.* **22**, 399. 100, 109

FEINGOLD (1951). *Rev. Mod. Phys.* **23**, 10. 100, 118

FERMI (1934). *Z. Phys.* **88**, 161. 93, 98

FERNBACH, SERBER and TAYLOR, T. B. (1949). *Phys. Rev.* **75**, 1352. 82

FESHBACH and WEISSKOPF (1949). *Phys. Rev.* **76**, 1550. 82

FIREMAN (1949). *Phys. Rev.* **75**, 323. 6

FLAMMERSFELD (1947). *Z. Naturf.* **2a**, 86. 4

FLÜGGE and KREBS (1944). *Naturwissenschaften*, **32**, 71. 86

FOLDY (1951). *Phys. Rev.* **83**, 397. 9

FOURNIER (1929). *C.R. Acad. Sci., Paris*, **188**, 1553. 19, 30

FOX, LEITH, WOUTERS and MACKENZIE (1950). American Report, AECD 2848. 82

FREEDMAN, WAGNER, JAFFEY and MAY (1951). American Report, D78. 99

FREEDMAN, WAGNER, JAFFEY and MAY (1952). American Report, D82. 99

GAMOW (1928). *Z. Phys.* **51**, 204. 28

GAMOW (1930). *Nature, Lond.*, **126**, 397. 34

GAMOW (1932). *Nature, Lond.*, **129**, 470. 34

GAMOW and HOUTERMANS (1928). *Z. Phys.* **52**, 496. 33

GAMOW and TELLER (1936). *Phys. Rev.* **49**, 895. 94, 98

GANS, HARKINS and NEWSON (1933). *Phys. Rev.* **44**, 310. 3

GAPON (1932). *Z. Phys.* **79**, 676. 26

GAPON (1933). *Z. Phys.* **81**, 419. 26

GARDNER, KNABLE and MOYER (1951). *Phys. Rev.* **83**, 1054. 146

GEIGER (1921). *Z. Phys.* **8**, 45. 33

GEIGER and NUTTALL (1911). *Phil. Mag.* **22**, 613. 27

GEIGER and NUTTALL (1912*a*). *Phil. Mag.* **23**, 439. 27

GEIGER and NUTTALL (1912*b*). *Phil. Mag.* **24**, 647. 27

GERJUOY (1951). *Phys. Rev.* **81**, 62. 106
GHIORSO, JAFFEY, ROBINSON and WEISSBOURD (1949). *American NNES*, **14**B, 16.8. *page* 34
GHIORSO, MEINKE and SEABORG (1948). *Phys. Rev.* **74**, 695. 87
GHIORSO, MEINKE and SEABORG (1949). *Phys. Rev.* **76**, 1414. 51
GHIORSO, THOMPSON, S. G., STREET and SEABORG (1951). *Phys. Rev.* **81**, 154. 42
GLENDENIN (1949). *Phys. Rev.* **75**, 337. 77
GLENDENIN and CORYELL (1950). *Phys. Rev.* **77**, 755. 77
GLENDENIN, CORYELL and EDWARDS (1949). *Phys. Rev.* **75**, 337. 77
GLUECKAUF (1948). *Proc. Phys. Soc.* **61**, 25. 67
GOLDIN, KNIGHT, MACKLIN, P. A. and MACKLIN, R. L. (1949). *Phys. Rev.* **76**, 336. 34
GOOD (1951). *Phys. Rev.* **81**, 1058. 4
GRACE, ALLEN, R. A., WEST and HALBAN (1951). *Proc. Phys. Soc.* A, **64**, 493. 35
GREEN and LIVESEY (1948). *Philos. Trans.* A, **241**, 323. 76
GUGGENHEIMER (1934*a*). *J. Physique*, **5**, 253. 26
GUGGENHEIMER (1934*b*). *J. Physique*, **5**, 475. 26
GUGGENHEIMER (1942). *Proc. Roy. Soc.* A, **181**, 169. 65
GURNEY and CONDON (1928). *Nature, Lond.*, **122**, 439. 28
GURNEY and CONDON (1929). *Phys. Rev.* **33**, 127. 28
HAEFNER (1951). *Rev. Mod. Phys.* **23**, 228. 15
HAHN, STRASSMANN and WALLING (1937). *Naturwissenschaften*, **25**, 189. 4
HALL, K. L. and TEMPLETON (1950). American Report, UCRL 957. 84
HANNA, G. C., HARVEY, B. G., MOSS and TUNNICLIFFE (1951). *Phys. Rev.* **81**, 466. 91
HANNA, G. C. and PONTECORVO (1949). *Phys. Rev.* **75**, 983. 7
HARKINS (1921*a*). *J. Amer. Chem. Soc.* **43**, 1038. 19
HARKINS (1921*b*). *Phil. Mag.* **42**, 305. 20
HARKINS (1933). *Proc. Nat. Acad. Sci., Wash.*, **19**, 307. 11
HARKINS (1949). *Phys. Rev.* **76**, 1538. 26
HARVEY, J. A. (1951). *Phys. Rev.* **81**, 353. 78
HAYDEN (1948). *Phys. Rev.* **74**, 650. 17, 22
HEMMENDINGER (1948). *Phys. Rev.* **73**, 806. 14
HEMMENDINGER (1949). *Phys. Rev.* **75**, 1267. 14
HENDERSON, G. H. and TURNBULL (1934). *Proc. Roy. Soc.* A, **145**, 582. 75
HESS and INGHRAM (1949). *Phys. Rev.* **76**, 1717. 4
v. HEVESY (1935). *Naturwissenschaften*, **23**, 583. 4
HEYDEN and WEFELMEIER (1938). *Naturwissenschaften*, **26**, 612. 4, 60
HILL (1948). *Proc. Camb. Phil. Soc.* **44**, 440. 33
HOWLAND, TEMPLETON and PERLMAN (1947). *Phys. Rev.* **71**, 552. 51
HUGHES, DABBS, CAHN and HALL, D. (1948). *Phys. Rev.* **73**, 111. 144
HUGHES and SHERMAN (1950). *Phys. Rev.* **78**, 632. 80
HUGHES, SPATZ and GOLDSTEIN (1949). *Phys. Rev.* **75**, 1781. 80
HULET, THOMPSON, S. G., GHIORSO and SEABORG (1951). American Report, UCRL 1224. 42
HYDE (1946*a*). American Report, AECD 2457. 34

HYDE (1946*b*). American Report, AECD 2648. *page* 34

HYDE, GHIORSO and SEABORG (1950). *Phys. Rev.* **77**, 765. 51

INGHRAM, BROWN, PATTERSON and HESS (1950). *Phys. Rev.* **80**, 916. 4, 14

INGHRAM, HAYDEN and HESS (1947*a*). *Phys. Rev.* **71**, 643. 17

INGHRAM, HAYDEN and HESS (1947*b*). *Phys. Rev.* **72**, 349. 4

INGHRAM, HAYDEN and HESS (1947*c*). *Phys. Rev.* **72**, 967. 4

INGHRAM, HESS and HAYDEN (1947). *Phys. Rev.* **72**, 1269. 17

INGHRAM and REYNOLDS, J. H. (1949). *Phys. Rev.* **76**, 1265. 5

ITOH (1940). *Proc. phys.-math. Soc. Japan*, **22**, 531. 105

IVANENKO and LEBEDER (1950). *J. Exp. Theor. Phys.* **20**, 11. 8

JEAN (1948). *C.R. Acad. Sci., Paris*, **226**, 2064. 8

JENKNER and BRODA (1949). *Nature, Lond.*, **164**, 412. 3

JENTSCHKE (1950). *Phys. Rev.* **77**, 98. 34, 86

JESSE and SADAUSKIS (1949). *Phys. Rev.* **75**, 1110. 2

JHA (1950). Ph.D. thesis. Edinburgh. 4

JONES, E. G. (1932). *Nature, Lond.*, **130**, 580. 26

JONES, J. W. and KOHMAN (1952). *Phys. Rev.* **85**, 941. 25

KALKSTEIN and LIBBY (1952). *Phys. Rev.* **85**, 368. 6

KAPLAN (1951). *Phys. Rev.* **81**, 962. 50

KARLIK (1948). *Acta phys. Austriaca*, **2**, 182. 34

KARLIK and BERNERT (1943*a*). *Naturwissenschaften*, **31**, 298. 84

KARLIK and BERNERT (1943*b*). *Naturwissenschaften*, **31**, 492. 86

KARLIK and BERNERT (1944). *Z. Phys.* **123**, 51. 84, 86

KARLIK and BERNERT (1946). *Naturwissenschaften*, **33**, 23. 84

KARRAKER and TEMPLETON (1950). American Report, UCRL 640. 51

KEVIL, LARSON and WANK (1944). *Amer. J. Sci.* **242**, 345. 73

KHLOPIN and ABIDOV (1941). *C.R. Acad. Sci. U.R.S.S.* **32**, 637. 74

KINSEY, HANNA, R. C. and VAN PATTER (1948). *Canad. J. Res.* A, **26**, 79. 76

KLINKENBERG (1945). *Physica*, **11**, 327. 78

KLINKENBERG (1951). *Physica*, **17**, 715. 60

KOCH, H. W., MCELHINNEY and GASTEIGER (1950). *Phys. Rev.* **77**, 329. 150

KOCH, J. and RASMUSSEN, E. (1950). *Phys. Rev.* **77**, 722. 80

KOFOED-HANSEN (1947). *Phys. Rev.* **71**, 451. 7

KOFOED-HANSEN (1951). *Phil. Mag.* **42**, 1448. 7

KOHMAN (1948). *Phys. Rev.* **73**, 16. 19

KOHMAN (1952). *Phys. Rev.* **85**, 530. 136

KONOPINSKI (1943). *Rev. Mod. Phys.* **15**, 209. 96, 100, 105

KONOPINSKI and UHLENBECK (1941). *Phys. Rev.* **60**, 308. 100

KOWARSKI (1950). *Phys. Rev.* **78**, 477. 18

KUNSTADTER, FLOYD, BORST and WEREMCHUK (1951). *Phys. Rev.* **83**, 235. 144

LANDÉ (1933*a*). *Phys. Rev.* **43**, 620. 26

LANDÉ (1933*b*). *Phys. Rev.* **43**, 624. 26

LAWSON (1951). *Phys. Rev.* **81**, 299. 6

LEININGER, SEGRÈ and SPIESS (1951). *Phys. Rev.* **82**, 334. 35

LELAND (1949*a*). *Phys. Rev.* **76**, 992. 15, 26

LELAND (1949*b*). *Phys. Rev.* **76**, 1722. 4

LEVINE, GHIORSO and SEABORG (1950). *Phys. Rev.* **77**, 296. *page* 10
LEVINGER, MEINERS, SAMPSON, SNELL and WILKINSON, R. G. (1950). *American NNES*, **9**, 63. 144
LEVY and PERLMAN (1951). American Report, UCRL 1532. 46
LIBBY (1934). *Phys. Rev.* **46**, 196. 3, 4
LIBBY (1939). *Phys. Rev.* **56**, 21. 4
LINDNER and PERLMAN (1948). *Phys. Rev.* **73**, 1124. 26
MACNAMARA, COLLINS and THODE (1950). *Phys. Rev.* **78**, 129. 77
MACNAMARA and THODE (1950). *Phys. Rev.* **80**, 471. 149
MAEDER and PREISWERK (1951). *Phys. Rev.* **84**, 595. 114
MAGNUSSON, THOMPSON, S. G. and SEABORG (1950). *Phys. Rev.* **78**, 363. 87
MANNING, ANDERSON and WATSON (1950). *Phys. Rev.* **78**, 417. 79
MARINSKY, GLENDENIN and CORYELL (1947). *J. Amer. Chem. Soc.* **69**, 2781. 17
MARSHAK (1942). *Phys. Rev.* **61**, 431. 7, 98
MARTELL and LIBBY (1950). *Phys. Rev.* **80**, 977. 4
MASSEY and MOHR (1936). *Proc. Roy. Soc.* A, **156**, 634. 15
MATTAUCH (1937). *Naturwissenschaften*, **25**, 189. 4
MAYER (1948). *Phys. Rev.* **74**, 235. 26, 145
MEINKE, GHIORSO and SEABORG (1951*a*). *Phys. Rev.* **81**, 782. 34
MEINKE, GHIORSO and SEABORG (1951*b*). American Report, UCRL 1141. 34
MEITNER (1926). *Naturwissenschaften*, **14**, 719. 6
MEITNER (1950). *Nature, Lond.*, **165**, 561. 77
MEYER, S. (1932). *S.B. Akad. Wiss. Wien*, II*a*, **141**, 71. 40
MICHEL (1950). *Nature, Lond.*, **166**, 654. 147
MOSZKOWSKI (1951). *Phys. Rev.* **82**, 35. 100, 110
MOTTA and BOYD (1948). *Phys. Rev.* **74**, 344. 17
MULHOLLAND and KOHMAN (1952). *Phys. Rev.* **85**, 144. 17
MURAKAWA and ROSS (1951). *Phys. Rev.* **83**, 1272. 79
NALDRETT and LIBBY (1948*a*). *Phys. Rev.* **73**, 487. 4
NALDRETT and LIBBY (1948*b*). *Phys. Rev.* **73**, 929. 4
NEUMANN, HOWLAND and PERLMAN (1950). *Phys. Rev.* **77**, 720. 46, 89
NEUMANN and PERLMAN (1950). *Phys. Rev.* **78**, 191. 51
NISHINA, YASAKI, KIMURA and IKAWA (1940). *Phys. Rev.* **58**, 660. 26
NORDSTRÖM (1950). *Z. Naturf.* **5***a*, 6. 83, 87
NORDSTRÖM (1951). *Ark. Fys.* **3**, 547. 62
PAPPAS (1951). *Phys. Rev.* **81**, 299. 77
PERLMAN, GHIORSO and SEABORG (1948). *Phys. Rev.* **74**, 1730. 34
PERLMAN, GHIORSO and SEABORG (1949). *Phys. Rev.* **75**, 1096. 34
PERLMAN, GHIORSO and SEABORG (1950). *Phys. Rev.* **77**, 26. 34, 87
PERLMAN and YPSILANTIS (1950). *Phys. Rev.* **79**, 30. 50
PETCH and JOHNS (1950). *Phys. Rev.* **80**, 478. 99
PETRZHAK and FLEROV (1940). *J. Physique U.S.S.R.* **3**, 275. 1, 148
PETSCHEK and MARSHAK (1952). *Phys. Rev.* **85**, 698. 103
PICCIOTTO (1949). *C.R. Acad. Sci., Paris*, **229**, 117. 2
PONTECORVO, KIRKWOOD and HANNA, G. C. (1949). *Phys. Rev.* **75**, 982. 7
PRESTON, M. A. (1946). *Phys. Rev.* **69**, 535. 50

PRESTON, M. A. (1947). *Phys. Rev.* **71**, 865. *pages* 28, 5

PRESTON, M. A. (1949). *Phys. Rev.* **75**, 90. 2

PRINGLE, STANDIL and ROULSTON (1950). *Phys. Rev.* **78**, 303.

PRINGLE, STANDIL, TAYLOR, H. W. and FRYER (1951). *Phys. Rev.* **84**, 1066.

PRYCE (1950). *Proc. Phys. Soc.* A, **63**, 692. 67, 12

RACAH (1932). *Nature, Lond.*, **129**, 723. 7

RASMUSSEN, J. O., REYNOLDS, F. L., THOMPSON, S. G. and GHIORSO (1950). *Phys. Rev.* **80**, 475. 71, 73

RASMUSSEN, J. O., THOMPSON, S. G. and GHIORSO (1951). American Report, UCRL 1473. 50, 73

RAYLEIGH (1933). *Nature, Lond.*, **131**, 724. 3

REDMAN and SAXON (1947). *Phys. Rev.* **72**, 570. 144

ROBERTS, MEYER, R. C. and WANG (1939). *Phys. Rev.* **55**, 510. 144

ROBSON (1950). *Phys. Rev.* **78**, 311. 6

ROBSON (1951). *Phys. Rev.* **83**, 349. 6

ROSE and JACKSON (1949). *Phys. Rev.* **76**, 1540. 115

ROSENBLUM (1929). *C.R. Acad. Sci., Paris*, **188**, 1401. 34

ROSENBLUM, GUILLOT and PEREY (1937). *C.R. Acad. Sci., Paris*, **204**, 175. 59

ROSENBLUM and VALADARES (1950). *C.R. Acad. Sci., Paris*, **230**, 384. 50

ROTBLAT (1943). British Report, Br 241. 149

RUSSELL (1923). *Phil. Mag.* **46**, 642. 11, 20

RUSSELL (1924*a*). *Phil. Mag.* **47**, 1121. 20

RUSSELL (1924*b*). *Phil. Mag.* **48**, 365. 20

RUTHERFORD (1924). *J. Franklin Inst.* **198**, 725. 26

SAHA, A. K. (1944). *Proc. Nat. Inst. Sci. India*, **10**, 373. 28, 32

SAHA, M. N. and SAHA, A. K. (1946). *Trans. Nat. Inst. Sci. India*, **2**, 193. 121

SARGENT (1933). *Proc. Roy. Soc.* A, **139**, 659. 93

SARGENT (1939). *Canad. J. Res.* A, **17**, 82. 93

SAWYER and WIEDENBECK (1949). *Phys. Rev.* **76**, 1535. 4

SAWYER and WIEDENBECK (1950). *Phys. Rev.* **79**, 490. 4

SCHINTLMEISTER (1936). *S.B. Akad. Wiss. Wien*, II*a*, **145**, 449. 75

SCHINTLMEISTER (1938). *Öst. Chem. Z.* **41**, 315. 86

SCHÜLER and KEYSTON (1931). *Z. Phys.* **70**, 1. 78

SCHÜLER and SCHMIDT (1934). *Z. Phys.* **92**, 148. 78

SCOTT and TITTERTON (1950). *Phil. Mag.* **41**, 918. 147

SEABORG (1952). *Phys. Rev.* **85**, 157. 150

SEABORG and PERLMAN (1948). *Rev. Mod. Phys.* **20**, 585. 34

SEGRÈ (1951). American Report, AECD 3149. 148

SEGRÈ and SEABORG (1941). *Phys. Rev.* **59**, 212. 26

SEIDLITZ, BLEULER and TENDAM (1949). *Phys. Rev.* **76**, 861. 123

SHERK (1949). *Phys. Rev.* **75**, 789. 8

SIEGBAHN and SLÄTIS (1947). *Ark. Mat. Astr. Fys.* **34**A, 15. 35

SINMA and YAMASAKI (1941). *Sci. Pap. Inst. phys. chem. Res. Tokyo*, **38**, 167. 81

SITTE (1948). *Acta phys. Austriaca*, **2**, 1. 141

SMYTHE and HEMMENDINGER (1937). *Phys. Rev.* **51**, 178. 4

SNELL, LEVINGER, MEINERS, SAMPSON and WILKINSON, R. G. (1947). *Phys. Rev.* **72**, 545. *page* 144
SNELL, NEDZEL, IBSER, LEVINGER, WILKINSON, R. G. and SAMPSON (1947). *Phys. Rev.* **72**, 541. 144
SNELL, PLEASANTON and McCORD (1950). *Phys. Rev.* **78**, 310. 6
SODDY (1909). *Phil. Mag.* **18**, 739. 83
SOKOLOW (1927). *Z. Phys.* **45**, 409. 40
SPIERS (1950). *Nature, Lond.*, **165**, 356. 4
SPIESS (1951). American Report, UCRL 1494. 35, 62
STANLEY and KATCOFF (1949). *J. Chem. Phys.* **17**, 653. 77, 143
STEHNEY (1950). American Report, AECD 3047. 144
STOUT (1949). *Phys. Rev.* **75**, 1107. 4
STRACHAN (1948). *Proc. Phys. Soc.* **61**, 481. 98
STRUTT (1908). *Proc. Roy. Soc.* A, **80**, 572. 73
STRUTT (1910). *Proc. Roy. Soc.* A, **83**, 298. 73
STUDIER and HYDE (1948). *Phys. Rev.* **74**, 591. 84
SUESS (1951). *Phys. Rev.* **81**, 1071. 124, 127
SUESS and JENSEN (1951). *Siegbahn presentation volume*, p. 589. Uppsala. 121
SUGARMAN (1947). *J. Chem. Phys.* **15**, 544. 144, 145
SUGARMAN and RICHTER (1948). *Phys. Rev.* **73**, 1411. 4
SULLIVAN (1949). *Trilinear Chart of Nuclear Species.* John Wiley and Sons. 34
SUN, CHARPIE, PECJAK and JENNINGS (1950). *Westinghouse Res. Lab. Sci. Paper* no. 1498. 144
SUN, PECJAK, JENNINGS, ALLEN, A. J. and NECHAJ (1951). *Phys. Rev.* **82**, 772. 73
SWINNE (1912). *Phys. Z.* **13**, 14. 28
SWINNE (1913). *Phys. Z.* **14**, 142. 28
TEMPLETON (1950). American Report, UCRL 616. 89
TEMPLETON and PERLMAN (1948). *Phys. Rev.* **73**, 1211. 51
THOMPSON, F. C. and ROWLANDS (1943). *Nature, Lond.*, **152**, 103. 4
THOMPSON, S. G. (1949). *Phys. Rev.* **76**, 319. 114
THOMPSON, S. G., GHIORSO, RASMUSSEN, J. O. and SEABORG (1949). *Phys. Rev.* **76**, 1406. 51, 72
THOMPSON, S. G., GHIORSO and SEABORG (1950). *Phys. Rev.* **80**, 781. 42
THOMPSON, S. G., STREET, GHIORSO and SEABORG (1950). *Phys. Rev.* **80**, 790. 42
TITTERTON (1951). *Nature, Lond.*, **168**, 590. 76
TOLLESTRUP, FOWLER and LAURITSEN (1949). *Phys. Rev.* **76**, 428. 14
TOMKINS, FRED and MEGGERS (1951). *Phys. Rev.* **84**, 168. 60
TOUSCHEK (1948). *Z. Phys.* **125**, 108. 10
TURNER (1940). *Phys. Rev.* **57**, 950. 19, 83
TURNER (1945). *Rev. Mod. Phys.* **17**, 292. 91, 149
VIGNERON (1947). *C.R. Acad. Sci., Paris*, **225**, 1067. 34, 86
WAPSTRA (1950). *Physica*, **16**, 33. 90
WAPSTRA (1951). *Physica*, **17**, 628. 70
WAPSTRA (1952). *Physica*, **18**, 83. 4, 71, 83, 135
WAY and WIGNER (1948). *Phys. Rev.* **73**, 1318. 145

Weaver (1950). *Phys. Rev.* **80**, 301. *page* 7
Wefelmeier (1937). *Z. Phys.* **107**, 332. 2
v. Weizsäcker (1935). *Z. Phys.* **96**, 431. 12
v. Weizsäcker (1937). *Phys. Z.* **38**, 623. 4, 14
Wheeler (1937). *Phys. Rev.* **52**, 1083. 15
Wheeler (1941). *Phys. Rev.* **59**, 27. 15
Whitehouse and Galbraith (1950). *Phil. Mag.* **41**, 429. 149
Whitehouse and Galbraith (1952). *Nature, Lond.*, **169**, 492. 150
Wigner (1939). *Phys. Rev.* **56**, 519. 112
Wigner and Feenberg (1941). *Rep. Progr. Phys.* **8**, 274. 121
Wilkinson, D. H. (1950). *Ionization Chambers and Counters*, p. 128. Cambridge University Press. 34
Winans (1947). *Phys. Rev.* **72**, 435. 26
Wolff (1921). *Phys. Z.* **22**, 171. 30

SUBJECT INDEX

Printed in the United States
By Bookmasters